Alfred Cardew Dixon

The elementary properties of the elliptic functions

With examples

Alfred Cardew Dixon

The elementary properties of the elliptic functions
With examples

ISBN/EAN: 9783742892461

Manufactured in Europe, USA, Canada, Australia, Japa

Cover: Foto ©berggeist007 / pixelio.de

Manufactured and distributed by brebook publishing software (www.brebook.com)

Alfred Cardew Dixon

The elementary properties of the elliptic functions

THE ELEMENTARY PROPERTIES

OF THE

ELLIPTIC FUNCTIONS

WITH EXAMPLES

BY

ALFRED CARDEW DIXON, M.A.

LATE FELLOW OF TRINITY COLLEGE, CAMBRIDGE; PROFESSOR OF MATHEMATICS
AT QUEEN'S COLLEGE, GALWAY

London
MACMILLAN AND CO.
AND NEW YORK
1894

All rights reserved

PREFACE.

THE object of this work is to supply the wants of those students who, for reasons connected with examinations or otherwise, wish to have a knowledge of "the elements of Elliptic Functions, not including the Theory of Transformations and the Theta Functions." It is right that I should acknowledge my obligations to the treatise of Professor Cayley and to the lectures of Dr. Glaisher, as well as to the authorities referred to from time to time. I am also greatly indebted to my brother, Mr. A. L. Dixon, Fellow of Merton College, Oxford, for his kind help in reading all the proofs and working through the examples, as also for his valuable suggestions.

A. C. DIXON.

DUBLIN, October, 1894.

CONTENTS.

CHAPTER I.
INTRODUCTION. DEFINITION OF ELLIPTIC FUNCTIONS, - 1

CHAPTER II.
FIRST DEDUCTIONS FROM THE DEFINITIONS. THE PERIODS. THE RELATED MODULI, - - - - - - 8

CHAPTER III.
ADDITION OF ARGUMENTS, - - - - - - 25

CHAPTER IV.
MULTIPLICATION AND DIVISION OF THE ARGUMENT, - - 38

CHAPTER V.
INTEGRATION, - - - - - - - - 46

CHAPTER VI.
ADDITION OF ARGUMENTS FOR THE FUNCTIONS E, Π, - 53

CHAPTER VII.
WEIERSTRASS' NOTATION, - - - - - - 63

CHAPTER VIII.
Degeneration of the Elliptic Functions, - - - 69

CHAPTER IX.
Differentiation with Respect to the Modulus, - - 73

CHAPTER X.
Applications, - - - - - - - - 82

APPENDIX A.
The Graphical Representation of Elliptic Functions, 129

APPENDIX B.
History of the Notation of the Subject, - - - 136

ELLIPTIC FUNCTIONS.

CHAPTER I.

INTRODUCTION. DEFINITION OF ELLIPTIC FUNCTIONS.

§ 1. In the earlier branches of mathematics functions are defined in various ways. Some are the results of the fundamental operations of algebra. $x+1$, $2x$, x^2 are such functions of x. Others are introduced by the inversion of those operations; such are $x-1$, $1/x$, $\sqrt{x}$; and others by conventional extensions of them, as $x^{-\frac{3}{4}}$, e^x. It is not easy to draw the line of distinction between the two last-named classes. Sometimes, again, geometrical constructions are used in the definition, as in the case of the trigonometrical functions.

§ 2. The elliptic functions cannot readily be defined in any of the foregoing ways; their fundamental property is that their differential coefficients can be expressed in a certain form, and as this is a somewhat new way of defining a function, we shall take one or two examples to show that it is as effective as any of those above mentioned.

§ 3. Let us define the exponential function by the equation

$$\frac{d}{du} \exp u = \exp u.$$

This equation tells us what addition is to be made to the value of $\exp u$ when a small change is made in that of u, and would therefore enable us gradually to find the value of the function for every value of the argument u, provided we knew one particular value to start with. Suppose then that when u has the value 0, $\exp u$ has the value 1, that is, $\exp 0 = 1$.

This equation combined with the former supplies a definition of the function $\exp u$.*

§ 4. From the foregoing definition we can deduce the properties of the function $\exp u$. First of all we can find an expression for $\exp(u+v)$.

Let $u+v=w$, and suppose w to be kept constant while u and v vary.

Then $$\frac{d}{du} \exp v = -\frac{d}{dv} \exp v = -\exp v.$$

Thus $$\exp u \cdot \frac{d}{du} \exp v + \exp v \cdot \frac{d}{du} \exp u = 0,$$

or $$\frac{d}{du}(\exp u \exp v) = 0.$$

Hence $\exp u \exp v$ is a constant as long as w is a constant, and has the same value whatever we may put for u and v so long as $u+v=w$.

---

* Compare the construction of trigonometrical tables, as explained in works on Trigonometry. The sine, tangent, etc., of every angle are found by adding the proper increments to those of an angle slightly less.

Put then $v=0$, $u=w$, and we have
$\exp u \exp v = \exp w \exp 0 = \exp(u+v)$, since $\exp 0 = 1$.

§ 5. We can also deduce the expansion of $\exp u$ in powers of u.

For $\qquad \dfrac{d}{du} \exp u = \exp u,$

so that $\qquad \dfrac{d^2}{du^2} \exp u = \dfrac{d}{du} \exp u = \exp u,$

and $\qquad \dfrac{d^r}{du^r} \exp u = \exp u,$

which $= 1$, when $u = 0$.

Thus Maclaurin's Theorem gives
$$\exp u = 1 + u + \frac{u^2}{2!} + \ldots + \frac{u^r}{r!} + \ldots,$$
the convergency of which may be established in the usual way.

§ 6. As another example, define the sine and cosine by the equation
$$\frac{d}{du} \sin u = \cos u, \ldots\ldots\ldots\ldots\ldots\ldots(1)$$
where $\qquad \cos^2 u + \sin^2 u = 1, \ldots\ldots\ldots\ldots\ldots\ldots(2)$
and $\qquad \sin 0 = 0, \ \cos 0 = 1.$

§ 7. Differentiating (2), we have
$$\cos u \frac{d}{du} \cos u + \sin u \cos u = 0,$$
whence $\qquad \dfrac{d}{du} \cos u = -\sin u, \ldots\ldots\ldots\ldots\ldots(3)$
as $\cos u$ is not zero in general.

§ 8. To find $\sin(u+v)$ and $\cos(u+v)$ put $u+v=w$, a constant, as before.

Consider a symmetrical function of u and v, such as $\sin u + \sin v$.

$$\frac{d}{du}(\sin u + \sin v) = \cos u - \cos v.$$

In the same way

$$\frac{d}{du}(\cos u + \cos v) = -\sin u + \sin v.$$

But $\cos^2 u + \sin^2 u = \cos^2 v + \sin^2 v$,

so that $(\cos u - \cos v)(\cos u + \cos v)$

$$= (-\sin u + \sin v)(\sin u + \sin v). \quad \ldots\ldots(4)$$

Hence $(\cos u + \cos v)\dfrac{d}{du}(\sin u + \sin v)$

$$= (\sin u + \sin v)\frac{d}{du}(\cos u + \cos v),$$

so that $\dfrac{\sin u + \sin v}{\cos u + \cos v} = $ a const. $= \dfrac{\sin(u+v)}{\cos(u+v)+1}, \ldots\ldots(5)$

putting w for u and 0 for v.

Then from (4) and (5)

$$\frac{-\sin u + \sin v}{\cos u - \cos v} = \text{a const. also,}$$

$$= \frac{\sin(u+v)}{1 - \cos(u+v)}.$$

And we find by solving

$$\sin(u+v) = \frac{\sin^2 u - \sin^2 v}{\sin u \cos v - \sin v \cos u}$$

$$= \sin u \cos v + \sin v \cos u \text{ by help of (2).}$$

Here again the functions may be expanded by Maclaurin's Theorem.

INTRODUCTION.

§ 9. The equations of definition are satisfied also if we change the signs of u and of $\sin u$. Thus

$$\sin(-u) = -\sin u,$$
$$\cos(-u) = \cos u.$$

The equations (1) and (2) are also satisfied if $\cos u$ is put for $\sin u$ and $-\sin u$ for $\cos u$. The initial values however are now different and a constant must be added to u. Call this constant ϖ.

Then
$$\sin(u+\varpi) = \cos u,$$
$$\cos(u+\varpi) = -\sin u,$$

if ϖ is such that $\sin \varpi = 1$, $\cos \varpi = 0$. Hence

$$\sin(u+2\varpi) = \cos(u+\varpi) = -\sin u,$$
$$\cos(u+2\varpi) = -\sin(u+\varpi) = -\cos u,$$
$$\sin(u+4\varpi) = -\sin(u+2\varpi) = \sin u,$$
$$\cos(u+4\varpi) = -\cos(u+2\varpi) = \cos u.$$

Hence the functions are unchanged when the argument u is increased by 4ϖ, that is to say, they are *periodic*.

§ 10. Again, writing ι for $\sqrt{-1}$,

$$\frac{d}{du}(\cos u + \iota \sin u) = \iota(\cos u + \iota \sin u),$$

or
$$\frac{d}{d\iota u}(\cos u + \iota \sin u) = \cos u + \iota \sin u,$$

and
$$\cos 0 + \iota \sin 0 = 1,$$

so that
$$\cos u + \iota \sin u = \exp \iota u.$$

This equation includes De Moivre's Theorem, and shows that $\exp u$ is also periodic, the period being $4\iota\varpi$.

These examples may be enough to show that functions which we know already can be defined in the way that was mentioned in § 2.

ELLIPTIC FUNCTIONS.

§ 11. Now the three elliptic functions $\operatorname{sn} u$, $\operatorname{cn} u$, $\operatorname{dn} u$ * are defined by the equations

$$\frac{d}{du} \operatorname{sn} u = \operatorname{cn} u \, \operatorname{dn} u,$$

$$\operatorname{cn}^2 u + \operatorname{sn}^2 u = 1, \dagger$$

$$\operatorname{dn}^2 u + k^2 \operatorname{sn}^2 u = 1,$$

$$\operatorname{sn} 0 = 0, \quad \operatorname{cn} 0 = \operatorname{dn} 0 = 1.$$

From these it follows at once that

$$\frac{d}{du} \operatorname{cn} u = - \operatorname{sn} u \, \operatorname{dn} u,$$

$$\frac{d}{du} \operatorname{dn} u = - k^2 \operatorname{sn} u \, \operatorname{cn} u.$$

The quantity k is a constant, called the *modulus*; u is called the *argument*.

§ 12. For different values of the modulus k (or, perhaps, rather of k^2, as the first power of k does not appear in the definition) there will be different values of the elliptic functions of any particular argument, in fact, $\operatorname{sn} u$, $\operatorname{cn} u$, $\operatorname{dn} u$ are really functions of two independent variables, and when it is desirable to call this fact to mind we shall write them

$$\operatorname{sn}(u, k), \quad \operatorname{cn}(u, k), \quad \operatorname{dn}(u, k).$$

We shall also use the following convenient and suggestive notation, invented by Dr. Glaisher:—

$$\operatorname{cn} u / \operatorname{dn} u = \operatorname{cd} u, \quad \operatorname{sn} u / \operatorname{cn} u = \operatorname{sc} u,$$

$$\operatorname{dn} u / \operatorname{cn} u = \operatorname{dc} u, \quad 1/\operatorname{sn} u = \operatorname{ns} u,$$

$$1/\operatorname{cn} u = \operatorname{nc} u, \text{ etc.}$$

It is usual to write k' for $(1-k^2)^{\frac{1}{2}}$, and k' is called the *complementary modulus*.

---

* Read s, n, u—c, n, u—d, n, u.
† Here and elsewhere $\operatorname{sn}^2 u$, etc., stand for $(\operatorname{sn} u)^2$, etc., as in Trigonometry.

INTRODUCTION. 7

The reader will not fail to notice the analogy between the two functions sn u and sin u, as also that between cos u and either cn u or dn u. (Compare §§ 74-75 below.)

EXAMPLES ON CHAPTER I.

1. Find the value of tan$(u+v)$ in terms of tan u and tan v from the equations

$$\frac{d}{du}\tan u = 1+\tan^2 u, \quad \tan 0 = 0.$$

2. Prove also that tan u is a periodic function of u, the period being twice that value of u for which tan u is infinite.

3. Find the value of sech$(u+v)$, given that

$$\frac{d}{du}\operatorname{sech} u = -\operatorname{sech} u \tanh u,$$

where $\operatorname{sech}^2 u + \tanh^2 u = 1$,
and that $\operatorname{sech} 0 = 1, \quad \tanh 0 = 0.$

4. Find the differential coefficients with respect to u of ns u, nc u, nd u, sc u, sd u, cs u, cd u, ds u, dc u.

Ans. $-\operatorname{cs} u\, \operatorname{ds} u, \quad \operatorname{sc} u\, \operatorname{dc} u, \quad k^2 \operatorname{sd} u\, \operatorname{cd} u, \quad \operatorname{nc} u\, \operatorname{dc} u,$
$\operatorname{nd} u\, \operatorname{cd} u, \quad -\operatorname{ns} u\, \operatorname{ds} u, \quad -k'^2 \operatorname{sd} u\, \operatorname{nd} u, \quad -\operatorname{cs} u\, \operatorname{ns} u,$
$k'^2 \operatorname{sc} u\, \operatorname{nc} u.$

5. Differentiate with respect to u
   (1) sn $u/(1+\operatorname{cn} u)$.   Ans. dn $u/(1+\operatorname{cn} u)$.
   (2) sn $u/(1+\operatorname{dn} u)$.   Ans. cn $u/(1+\operatorname{dn} u)$.
   (3) cn $u/(1+\operatorname{sn} u)$.   Ans. $-\operatorname{dn} u/(1+\operatorname{sn} u)$.
   (4) dn $u/(1+k\operatorname{sn} u)$.   Ans. $-k \operatorname{cn} u/(1+k\operatorname{sn} u)$.
   (5) arcsin sn u.   Ans. dn u.
   (6) sn $u/(\operatorname{dn} u - \operatorname{cn} u)$. Ans. $1/(\operatorname{cn} u - \operatorname{dn} u)$.

CHAPTER II.

FIRST DEDUCTIONS FROM THE DEFINITIONS. THE PERIODS. THE RELATED MODULI.

§ 13. It follows from the foregoing definitions that if a function S or $S(v)$ of a variable v satisfies the equation

$$\frac{dS}{dv} = CD, \dots\dots\dots\dots\dots(1)$$

where C and D are other functions of v connected with S by the equations

$$C^2 + S^2 = 1, \dots\dots\dots\dots\dots(2)$$
$$D^2 + \lambda^2 S^2 = 1; \dots\dots\dots\dots\dots(3)$$

then
$$S = \operatorname{sn}(v+a, \lambda), \dots\dots\dots\dots\dots(4)$$
$$C = \operatorname{cn}(v+a, \lambda), \dots\dots\dots\dots\dots(5)$$
$$D = \operatorname{dn}(v+a, \lambda), \dots\dots\dots\dots\dots(6)$$

where a is such a constant that

$$\operatorname{sn}(a, \lambda) = S(0), \dots\dots\dots\dots\dots(7)$$
$$\operatorname{cn}(a, \lambda) = C(0), \dots\dots\dots\dots\dots(8)$$
$$\operatorname{dn}(a, \lambda) = D(0), \dots\dots\dots\dots\dots(9)$$

these last equations being clearly consistent.

§ 14. Now, in the first place, the foregoing conditions hold if we put
$$S = -\operatorname{sn} u,\ C = \operatorname{cn} u,\ D = \operatorname{dn} u,\ \lambda = k,\ v = -u,\ a = 0;$$
and thus
$$\left.\begin{array}{l}\operatorname{sn}(-u) = -\operatorname{sn} u, \\ \operatorname{cn}(-u) = \operatorname{cn} u, \\ \operatorname{dn}(-u) = \operatorname{dn} u,\end{array}\right\} \quad \ldots\ldots\ldots\ldots\ldots(10)$$
or cn and dn are even functions, and sn is an odd function.

§ 15. We have also
$$\frac{d}{du} \operatorname{sc} u = (\operatorname{cn}^2 u\ \operatorname{dn} u + \operatorname{sn}^2 u\ \operatorname{dn} u)/\operatorname{cn}^2 u$$
$$= \operatorname{dn} u / \operatorname{cn}^2 u = \operatorname{dc} u\ \operatorname{nc} u,$$
and in the same way
$$\frac{d}{du} \operatorname{nc} u = \operatorname{sc} u\ \operatorname{dc} u,$$
$$\frac{d}{du} \operatorname{dc} u = k'^2 \operatorname{sc} u\ \operatorname{nc} u,$$
$$\frac{d}{du} \operatorname{cs} u = -\operatorname{ds} u\ \operatorname{ns} u,$$
$$\frac{d}{du} \operatorname{ns} u = -\operatorname{cs} u\ \operatorname{ds} u,$$
$$\frac{d}{du} \operatorname{ds} u = -\operatorname{cs} u\ \operatorname{ns} u,$$
$$\frac{d}{du} \operatorname{sd} u = \operatorname{cd} u\ \operatorname{nd} u,$$
$$\frac{d}{du} \operatorname{cd} u = -k'^2 \operatorname{sd} u\ \operatorname{nd} u,$$
$$\frac{d}{du} \operatorname{nd} u = k^2 \operatorname{sd} u\ \operatorname{cd} u. \quad \ldots\ldots\ldots\ldots(11)$$

By integrating these equations we shall deduce several important theorems.

§ 16. Take for instance
$$\frac{d}{du}\operatorname{cd} u = -k^2 \operatorname{sd} u \operatorname{nd} u.$$

We have $\operatorname{cn}^2 u + \operatorname{sn}^2 u = 1$,
$\operatorname{dn}^2 u + k^2 \operatorname{sn}^2 u = 1$;

and dividing by $\operatorname{dn}^2 u$,
$$\operatorname{cd}^2 u + \operatorname{sd}^2 u = \operatorname{nd}^2 u,$$
$$1 + k^2 \operatorname{sd}^2 u = \operatorname{nd}^2 u.$$

Hence $k^2 \operatorname{sd}^2 u + \operatorname{cd}^2 u = 1$, by elimination of $\operatorname{nd}^2 u$,
and $k'^2 \operatorname{nd}^2 u + k^2 \operatorname{cd}^2 u = 1$, by elimination of $\operatorname{sd}^2 u$.

In the equations (1) ... (6) of this chapter we may therefore put
$$S = \operatorname{cd} u, \quad C = -k' \operatorname{sd} u, \quad D = k' \operatorname{nd} u, \quad \lambda = k, \quad v = u.$$

The value of a is such that
$$\operatorname{sn} a = 1, \quad \operatorname{cn} a = 0, \quad \operatorname{dn} a = k'.$$

Let us write K for this value of a; then we have
$$\left. \begin{aligned} \operatorname{sn}(u+K) &= \operatorname{cd} u, \\ \operatorname{cn}(u+K) &= -k' \operatorname{sd} u, \\ \operatorname{dn}(u+K) &= k' \operatorname{nd} u. \end{aligned} \right\} \quad \ldots\ldots\ldots\ldots\ldots(12)$$

§ 17. From these it further follows that
$\operatorname{sn}(u+2K) = \operatorname{cd}(u+K) = -k' \operatorname{sd} u \div k' \operatorname{nd} u = -\operatorname{sn} u$,
$\operatorname{cn}(u+2K) = -k' \operatorname{sd}(u+K) = -k' \operatorname{cd} u \div k' \operatorname{nd} u = -\operatorname{cn} u$,
$\operatorname{dn}(u+2K) = k' \div \operatorname{dn}(u+K) = \operatorname{dn} u$.

Also $\operatorname{sn}(u+3K) = -\operatorname{sn}(u+K) = -\operatorname{cd} u$,
$\operatorname{cn}(u+3K) = k' \operatorname{sd} u$,
$\operatorname{dn}(u+3K) = k' \operatorname{nd} u$,
$\operatorname{sn}(u+4K) = -\operatorname{sn}(u+2K) = \operatorname{sn} u$,
$\operatorname{cn}(u+4K) = \operatorname{cn} u$,
$\operatorname{dn}(u+4K) = \operatorname{dn} u$.

THE PERIODS.

Again, $\quad \operatorname{sn}(K-u) = \operatorname{cd}(-u) = \operatorname{cd} u,$
$\operatorname{cn}(K-u) = k' \operatorname{sd} u,$
$\operatorname{dn}(K-u) = k' \operatorname{nd} u.$

Thus the function $\operatorname{dn} u$ is unaltered when its argument is increased by $2K$; $\operatorname{sn} u$ and $\operatorname{cn} u$ are unaltered when the argument is increased by $4K$, that is to say the functions are periodic.

§ 18. Take now the equation
$$\frac{d}{du} \operatorname{ns} u = -\operatorname{cs} u \operatorname{ds} u,$$
where
$$-\operatorname{cs}^2 u + \operatorname{ns}^2 u = 1,$$
$$-\operatorname{ds}^2 u + \operatorname{ns}^2 u = k^2.$$

Here we may write
$$S = \frac{1}{k} \operatorname{ns} u, \quad C = \frac{\iota}{k} \operatorname{ds} u, \quad D = \iota \operatorname{cs} u, \quad \lambda = k, \quad v = u,$$

but $\operatorname{sn} a$, $\operatorname{cn} a$, $\operatorname{dn} a$ are all infinite. We have, however,
$$\operatorname{cs} a = \iota, \quad \operatorname{ds} a = \iota k.$$

Let this value of a be called L for the time being.

Then
$$\left.\begin{array}{l} \operatorname{sn}(u+L) = \dfrac{1}{k} \operatorname{ns} u, \\[4pt] \operatorname{cn}(u+L) = \dfrac{\iota}{k} \operatorname{ds} u, \\ \operatorname{dn}(u+L) = \iota \operatorname{cs} u, \\ \operatorname{sn}(u+2L) = \operatorname{sn} u, \\ \operatorname{cn}(u+2L) = -\operatorname{cn} u, \\ \operatorname{dn}(u+2L) = -\operatorname{dn} u, \\ \operatorname{cn}(u+3L) = -\dfrac{\iota}{k} \operatorname{ds} u, \\ \operatorname{dn}(u+3L) = -\iota \operatorname{cs} u, \\ \operatorname{cn}(u+4L) = \operatorname{cn} u, \\ \operatorname{dn}(u+4L) = \operatorname{dn} u. \end{array}\right\} \quad \ldots\ldots\ldots\ldots(13)$$

12   ELLIPTIC FUNCTIONS.

§ 19. Also
$$\left.\begin{aligned}
&\operatorname{sn}(u+K+L) = \frac{1}{k}\operatorname{ns}(u+K) = \frac{1}{k}\operatorname{dc} u, \\
&\operatorname{cn}(u+K+L) = \frac{\iota}{k}\operatorname{ds}(u+K) = \frac{\iota k'}{k}\operatorname{nc} u, \\
&\operatorname{dn}(u+K+L) = \iota\operatorname{cs}(u+K) = -\iota k'\operatorname{sc} u, \\
&\operatorname{sn}(u+2K+2L) = -\operatorname{sn} u, \\
&\operatorname{cn}(u+2K+2L) = \operatorname{cn} u, \\
&\operatorname{dn}(u+2K+2L) = -\operatorname{dn} u.
\end{aligned}\right\} \ldots(14)$$

§ 20. Hence

sn u has a period $\quad 2L$ as well as $4K$,
cn u has a period $2K+2L$ as well as $4K$,
dn u has a period $\quad 4L$ as well as $2K$.

We may also notice that
$$\operatorname{sn}(K+L) = \frac{1}{k'}, \quad \operatorname{cn}(K+L) = \frac{\iota k'}{k}, \quad \operatorname{dn}(K+L) = 0.$$

THE COMPLEMENTARY MODULUS.

§ 21. Now consider the first equation of the system (11).
$$\frac{d}{du}\operatorname{sc} u = \operatorname{dc} u \operatorname{nc} u,$$
where
$$\operatorname{nc}^2 u - \operatorname{sc}^2 u = 1,$$
$$\operatorname{dc}^2 u - k'^2 \operatorname{sc}^2 u = 1.$$

Hence we may put
$$S = \iota \operatorname{sc} u, \quad C = \operatorname{nc} u, \quad D = \operatorname{dc} u, \quad v = \iota u, \quad \lambda = k',$$
in the equations (4), (5), (6); and as
$$S(0) = 0, \quad C(0) = D(0) = 1,$$
we have $\quad a = 0.$

THE COMPLEMENTARY MODULUS.

Thus
$$\left.\begin{array}{l}\operatorname{sn}(\iota u, k') = \iota \operatorname{sc}(u, k), \\ \operatorname{cn}(\iota u, k') = \operatorname{nc}(u, k), \\ \operatorname{dn}(\iota u, k') = \operatorname{dc}(u, k).\end{array}\right\}\ldots\ldots\ldots(15)$$

These equations are of great importance. They embody what is called *Jacobi's Imaginary Transformation* and enable us to express elliptic functions of purely imaginary arguments by means of those of real arguments with a different modulus.

§ 22. In the equations (15) put L for u.

Then
$$\operatorname{sn}(\iota L, k') = \iota \operatorname{sc}(L, k) = 1,$$
$$\operatorname{cn}(\iota L, k') = 0,$$
$$\operatorname{dn}(\iota L, k') = k.$$

Thus ιL stands to k' in the same relation as K to k, and we are naturally led to write

$$\iota L = K', \quad L = -\iota K'.$$

Thus if m and n are any two whole numbers

$$\left.\begin{array}{l}\operatorname{sn}(u + 2mK + 2n\iota K') = (-1)^m \operatorname{sn} u, \\ \operatorname{cn}(u + 2mK + 2n\iota K') = (-1)^{m+n} \operatorname{cn} u, \\ \operatorname{dn}(u + 2mK + 2n\iota K') = (-1)^n \operatorname{dn} u.\end{array}\right\}\ldots..(16)$$

We have then the following scheme for the values of sn, cn, dn, of $u + mK + n\iota K'$, m and n being integers:

	$m \equiv 0$,	$m \equiv 1$,	$m \equiv 2$,	$m \equiv 3$.
$n \equiv 0$	sn u,	cd u,	$-$sn u,	$-$cd u.
	cn u,	$-k'$sd u,	$-$cn u,	k'sd u.
	dn u,	k'nd u,	dn u,	k'nd u.
$n \equiv 1$	(ns u)/k,	(dc u)/k,	$-$(ns u)/k,	$-$(dc u)/k.
	$-\iota$(ds u)/k,	$-\iota k'$(nc u)/k,	ι(ds u)/k,	$\iota k'$(nc u)/k.
	$-\iota$ cs u,	$\iota k'$sc u,	$-\iota$ cs u,	$\iota k'$sc u.

ELLIPTIC FUNCTIONS.

	$m \equiv 0,$	$m \equiv 1,$	$m \equiv 2,$	$m \equiv 3.$
	sn u,	cd u,	$-$sn u,	$-$cd u.
$n \equiv 2$	$-$cn u,	k'sd u,	cn u,	$-k'$sd u.
	$-$dn u,	$-k'$nd u,	$-$dn u,	$-k'$nd u.
	(ns u)/k,	(dc u)/k,	$-$(ns u)/k,	$-$(dc u)/k.
$n \equiv 3$	ι(ds u)/k,	$\iota k'$(nc u)/k,	$-\iota$(ds u)/k,	$-\iota k'$(nc u)/k.
	ι cs u,	$-\iota k'$sc u,	ι cs u,	$-\iota k'$sc u.

the modulus in the congruences being 4.

§ 23. These equations show that a knowledge of the values of sn u, cn u, dn u does not enable us to fix the value of u, and that accordingly the value of K is not perfectly defined since we have only assigned the conditions

$$\text{sn } K = 1, \quad \text{cn } K = 0, \quad \text{dn } K = k'.$$

Writing x for sn u we have

$$\text{cn } u = (1-x^2)^{\frac{1}{2}}, \text{ dn } u = (1-k^2x^2)^{\frac{1}{2}},$$

$$\frac{dx}{du} = (1-x^2)^{\frac{1}{2}}(1-k^2x^2)^{\frac{1}{2}}.$$

Hence $$u = \int_0^x (1-\xi^2)^{-\frac{1}{2}}(1-k^2\xi^2)^{-\frac{1}{2}} d\xi,$$

the lower limit being 0 because u and x vanish together.

Thus $$K = \int_0^1 (1-\xi^2)^{-\frac{1}{2}}(1-k^2\xi^2)^{-\frac{1}{2}} d\xi.$$

This is a function of k only. The variable ξ will be supposed in the integration to pass continuously from 0 to 1 through all intermediate real values and those only, and the initial value of the subject of integration will be supposed to be unity and positive. There is now no ambiguity in the value of K so long as k^2 is less than 1. Also with the same provision K is a purely real positive quantity as every element in the integration is so.

Further, k' is to be the positive value of $(1-k^2)^{\frac{1}{2}}$, for $\operatorname{dn} u$ does not change sign within the limits of integration and $k' = \operatorname{dn} K$.

§ 24. Again, so long as k'^2 is less than 1, K' is also a purely real positive quantity.

Thus for values of the modulus between 0 and 1 the periods $4K$ and $4\iota K'$ are the one real, the other purely imaginary.

We shall now show how to reduce elliptic functions in which the square of the modulus is real, but not a positive proper fraction, to others in which the modulus lies between 0 and 1.

§ 25. We have $\dfrac{d}{du}\operatorname{sn} u = \operatorname{cn} u\, \operatorname{dn} u$,

$$\operatorname{cn}^2 u + \operatorname{sn}^2 u = 1,$$
$$\operatorname{dn}^2 u + k^2 \operatorname{sn}^2 u = 1,$$

and we may put
$$C = \operatorname{dn} u, \quad D = \operatorname{cn} u,$$
provided we have
$$S = k \operatorname{sn} u, \quad \lambda = 1/k, \quad v = ku.$$

Furthermore $a = 0$.

Thus
$$\left.\begin{array}{l}\operatorname{sn}(ku, 1/k) = k\operatorname{sn}(u, k),\\ \operatorname{cn}(ku, 1/k) = \operatorname{dn}(u, k),\\ \operatorname{dn}(ku, 1/k) = \operatorname{cn}(u, k).\end{array}\right\}\dots\dots\dots\dots(17)$$

The equations (17) enable us to reduce the case of a modulus numerically greater than unity to that of one less than unity.

§ 26. From the equations (15) and (17) we deduce
$$\begin{aligned}
\operatorname{sn}(\iota k'u, 1/k') &= k'\operatorname{sn}(\iota u, k') = \iota k'\operatorname{sc}(u, k), \\
\operatorname{cn}(\iota k'u, 1/k') &= \operatorname{dn}(\iota u, k') = \operatorname{dc}(u, k), \\
\operatorname{dn}(\iota k'u, 1/k') &= \operatorname{cn}(\iota u, k') = \operatorname{nc}(u, k),
\end{aligned} \right\} \dots(18)$$

and also, since $\iota k'/k$ is the modulus complementary to $1/k$,
$$\begin{aligned}
\operatorname{sn}(\iota ku, \iota k'/k) &= \iota \operatorname{sc}(ku, 1/k) = \iota k\operatorname{sd}(u, k), \\
\operatorname{cn}(\iota ku, \iota k'/k) &= \operatorname{nc}(ku, 1/k) = \operatorname{nd}(u, k), \\
\operatorname{dn}(\iota ku, \iota k'/k) &= \operatorname{dc}(ku, 1/k) = \operatorname{cd}(u, k),
\end{aligned} \right\} \dots(19)$$

and from (19) by help of (15)
$$\begin{aligned}
\operatorname{sn}(k'u, \iota k/k') &= -\iota k'\operatorname{sd}(\iota u, k') = k'\operatorname{sd}(u, k), \\
\operatorname{cn}(k'u, \iota k/k') &= \operatorname{nd}(\iota u, k') = \operatorname{cd}(u, k), \\
\operatorname{dn}(k'u, \iota k/k') &= \operatorname{cd}(\iota u, k') = \operatorname{nd}(u, k),
\end{aligned} \right\} \dots(20)$$

§ 27. The quantities corresponding to K, $\iota K'$, the quarter-periods, are given in the following table for the group of six *related moduli*:—

Modulus.	First Quarter-period.	Second Quarter-period.
k,	K,	$\iota K'$,
k',	K',	ιK,
$1/k$,	$k(K - \iota K')$,	$\iota k K'$,
$1/k'$,	$k'(K' - \iota K)$,	$\iota k' K$,
$\iota k'/k$,	kK',	$k(K' + \iota K)$,
$\iota k/k'$,	$k'K$,	$k'(K + \iota K')$,

the distinction being that $\operatorname{sn} = 1$ and $\operatorname{dn} =$ the complementary modulus for the first quarter-period, and that for the second sn, cn, dn are infinite and proportional to ι, 1 and the modulus.

§ 28. We can prove that if the modulus is a real proper fraction the elliptic functions of a real argument are real.

For as sn u increases from 0 to 1, while cn u decreases from 1 to 0, and dn u from 1 to k', the argument u increases continuously from 0 to K, so that for any value of u between 0 and K, sn u, cn u have real values between 0 and 1, dn u has a real value between k' and 1.

Also we see from §§ 14, 17 that
$$\text{sn}(2K-u) = \text{sn } u,$$
$$\text{cn}(2K-u) = -\text{cn } u,$$
$$\text{dn}(2K-u) = \text{dn } u,$$
so that when u lies between K and $2K$.

sn u is real and between 0 and 1,
cn u „   „   0 and -1,
dn u „   „   1 and k'.

Again,
$$\text{sn}(-u) = -\text{sn } u,$$
$$\text{cn}(-u) = \text{cn } u,$$
$$\text{dn}(-u) = \text{dn } u,$$
so that sn u, cn u, dn u are also real for values of u between 0 and $-2K$.

Also $\quad\text{sn}(u+4K) = \text{sn } u$, etc.,
so that, as any real quantity can be made up by adding a positive or negative multiple of $4K$ to a quantity between $\pm 2K$, sn u, cn u, dn u are all real if u is real. They are also real if u is a complex quantity whose imaginary part is a multiple of $2\iota K'$, for
$$\text{sn}(u+2\iota K') = \text{sn } u,$$
$$\text{cn}(u+2\iota K') = -\text{cn } u,$$
$$\text{dn}(u+2\iota K') = -\text{dn } u.$$

§ 29. Further, when the imaginary part of u is $\iota K'$, or an odd multiple of it,

sn u is real,
cn u and dn u are purely imaginary,

D. E. F.   B

18   ELLIPTIC FUNCTIONS.

for
$$\operatorname{sn}(u+\iota K') = 1/k\operatorname{sn} u,$$
$$\operatorname{cn}(u+\iota K') = -\iota\operatorname{dn} u/k\operatorname{sn} u,$$
$$\operatorname{dn}(u+\iota K') = -\iota\operatorname{cn} u/\operatorname{sn} u.$$

Again, since
$$\operatorname{sn}(\iota u, k) = \iota\operatorname{sc}(u, k'),$$
$$\operatorname{cn}(\iota u, k) = \operatorname{nc}(u, k'),$$
$$\operatorname{dn}(\iota u, k) = \operatorname{dc}(u, k'),$$

it follows that for a purely imaginary argument or a complex argument whose real part is a multiple of $2K$

sn is purely imaginary,
cn and dn are real.

Also, for a complex argument whose real part is an odd multiple of K

sn and dn are real,
cn is purely imaginary,

for $\operatorname{sn}(K+\iota u, k) = \operatorname{cd}(\iota u, k) = \operatorname{nd}(u, k'),$
$\operatorname{cn}(K+\iota u, k) = -k'\operatorname{sd}(\iota u, k) = -\iota k'\operatorname{sd}(u, k'),$
$\operatorname{dn}(K+\iota u, k) = k'\operatorname{nd}(\iota u, k) = k'\operatorname{cd}(u, k').$

§ 30. It is to be noticed that one of the periods at least is always imaginary or complex, and it may be proved that their ratio cannot be purely real.

For let ω_1 and ω_2 be two periods of a function $\phi(u)$ so that
$$\phi(u) = \phi(u+\omega_1) = \phi(u+\omega_2) = \phi(u+m\omega_1+n\omega_2),$$
m and n being any integers. Also let ω_1/ω_2 be real.

Two cases arise. If ω_1 and ω_2 have a common measure ω let
$$\omega_1 = p\omega, \quad \omega_2 = q\omega,$$
p and q being two integers prime to each other.

Then integral values of m and n can be found such that
$$mp+nq = 1,$$
so that
$$\phi(u+\omega) = \phi(u),$$
and the two periods ω_1, ω_2 reduce to one, ω.

§ 31. But if, on the other hand, ω_1 and ω_2 are incommensurable we can prove that $m\omega_1 + n\omega_2$ may be made smaller than any assignable finite quantity.

For let $\lambda\omega_2$ be the nearest multiple of ω_2 to ω_1; then

$$\omega_1 \sim \lambda\omega_2 (=\omega_3, \text{ say}) \text{ is less than } \tfrac{1}{2}\omega_2.$$

Let $\mu\omega_3$ be the nearest multiple of ω_3 to ω_2; then

$$\omega_2 \sim \mu\omega_3, \text{ or } \omega_4, \text{ is less than } \tfrac{1}{2}\omega_3,$$

and so on. Then

$$\omega_{2+r} \text{ is less than } \frac{1}{2^r}\omega_2,$$

which can be made smaller than any assignable finite quantity by taking r great enough. Also each of the quantities $\omega_3, \omega_4, \ldots$, is of the form $m\omega_1 + n\omega_2$, so that the statement is proved.

In this case then if $\phi(u + m\omega_1 + n\omega_2) = \phi(u)$, the value of the function is repeated at indefinitely short intervals, and the function must be either a constant or have an infinite number of values for each value of its argument.

§ 32. It may be proved that the same kind of consequences will follow if a function is supposed to have three periods whose ratios are complex.

We shall represent the argument of the function on Argand's diagram, in which the point P whose coordinates are (x, y) referred to rectangular axes OX, OY, represents the complex quantity $x + \iota y$. The statement that a straight line AB is a period will be understood to mean that if from any point P a line is drawn parallel to AB and equal to any multiple of it the value of the function is the same at the two ends of the line.

Now let OA, OB be two periods. Join AB. Through O, A, B draw lines parallel to AB, BO, OA respectively. Through their intersections draw other lines in the same directions and continue the process till the whole plane is covered with a network of triangles, each equal in all respects to the triangle OAB. Then any line joining two vertices of triangles of the system is a period, since each side of any triangle is one.

The triangles can be combined in pairs into parallelograms, all exactly alike, and similarly situated, and the values of the function at points similarly situated in different parallelograms will be the same. Such a parallelogram is called the 'parallelogram of the periods.'

Suppose, however, that there is a third period OC; then C must fall within or on the boundary of one triangle of the network. If it fall at an angular point then OC is not a new period, but is only a combination of OA and OB. If it fall on a side of a triangle, say DE, then DC and CE must be periods, and their ratio is real, since they are in the same direction; thus this case reduces to the one already discussed.

If C fall within a triangle, say DEF, then CD, CE, CF are all periods. Let G be the point similarly situated within the triangle OAB, then OG, AG, BG are all periods being respectively equal to CD, CE, CF in some order. Any of the triangles OBG, BAG, AOG may now be taken as the foundation of another network covering the whole plane, and since there is still a third period, we can again find a point within the fundamental triangle with which to carry on the same process. We can prove that ultimately either the point will fall on the boundary of one of the triangles, which case has been discussed above, or a period can be found shorter than any assigned finite straight line.

We shall form each triangle from the one before it as follows. Let Oab be a triangle of the series, and g the point found within it. Let $Oa \geqslant Ob$. Then we take Obg as the next triangle of the series.

IMPOSSIBILITY OF THREE PERIODS. 21

Let ϵ be any finite length, then we shall prove that a period can be found shorter than ϵ. Suppose that none such can be found among the sides of such triangles as ABG, ..., abg, ..., which have not O for a vertex.

The angle Oab is always acute, and can never be greater than $\tfrac{1}{2}\pi - \beta$ where β is some finite acute angle. For if there is no such limit, and Oab can be made to approach $\pi/2$ without limit, then since $Oba \geqslant Oab$, aOb can be diminished without limit, and therefore ab can be made less than ϵ.

If Oh is drawn perpendicular to ab and g falls within the triangle Ohb then $Og < Ob$.

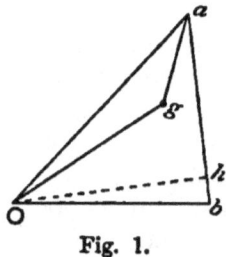

Fig. 1.

If not, we have

$$Oa - Og = ag \sin \tfrac{1}{2}(Oga - Oag) \div \cos \tfrac{1}{2}aOg > \epsilon \sin \tfrac{1}{2}\beta,$$

for $\quad ag \geqslant \epsilon, \quad Oga > Oha > \tfrac{1}{2}\pi, \quad Oag < \tfrac{1}{2}\pi - \beta.$

Thus Og is less than Oa by a finite quantity, and if $Og > Ob$ it will be reduced by a finite quantity at the next step and so on, until after a finite number of steps we have a triangle in which Ob is the greater side. We can then replace Ob by a line which is less by at least $\epsilon \sin \tfrac{1}{2}\beta$, and carry on the process, reducing this line again in the same way.

Let μ be the greatest integer in $Ob \div \epsilon \sin \tfrac{1}{2}\beta$. Then after μ stages at most the shorter side Ob of the triangle Oab will be replaced by a line less than

$\epsilon \sin \frac{1}{2}\beta$, and therefore less than ϵ. Each of these μ stages will consist of a finite number of steps by which the originally greater side of the triangle is gradually diminished till it becomes the less, followed by another step in which that which was the less originally is itself diminished.

It is proved then that if there are three periods ω_1, ω_2, ω_3, either they are not independent but satisfy an identity of the form $l\omega_1 + m\omega_2 + n\omega_3 = 0$ with integral coefficients, or else a period can be found whose *modulus* is smaller than any assignable finite quantity, so that the function has an infinite number of values for any single value of its argument. It might of course be a constant.

EXAMPLES ON CHAPTER II.

1. Prove that each of the twelve functions $\operatorname{sn} u$, $\operatorname{cn} u$, $\operatorname{ns} u$, ..., can be expressed as a multiple of the sn of an integral linear function of u with one of the six related moduli, in two ways, *e.g.*
$$\operatorname{dn}(u, k) = k'\operatorname{sn}(K' - \iota K - \iota u, k')$$
$$= \operatorname{sn}(k'K' - \iota k'K - \iota k'u, 1/k').$$

2. What are the periods of the functions $\operatorname{sc} u$, $\operatorname{dc} u$, $\operatorname{ds} u$, $\dfrac{\operatorname{sn} u}{1 + \operatorname{cn} u}$, $\dfrac{\operatorname{cn} u}{1 + \operatorname{sn} u}$, $\operatorname{sn} u \operatorname{cd} u$, $\operatorname{sn}^2 u$, $\dfrac{\operatorname{sn} u}{1 + k \operatorname{sn}^2 u}$?

3. Putting S for $\operatorname{sn} u \operatorname{sn}(u + K)$, verify that
$$\frac{dS}{du} = \frac{1}{k^2}\{\operatorname{dn}^2 u - \operatorname{dn}^2(u + K)\}, \quad \ldots\ldots\ldots\ldots(1)$$
$$\{\operatorname{dn} u + \operatorname{dn}(u + K)\}^2 + k^4 S^2 = (1 + k')^2, \quad \ldots\ldots(2)$$
$$\{\operatorname{dn} u - \operatorname{dn}(u + K)\}^2 + k^4 S^2 = (1 - k')^2. \quad \ldots\ldots(3)$$

Deduce that
$$(1 + k')S = \operatorname{sn}\left\{u(1 + k'), \frac{1 - k'}{1 + k'}\right\},$$

EXAMPLES II.

and find the values of

$$\operatorname{cn}\left\{u(1+k'), \frac{1-k'}{1+k'}\right\} \quad \text{and} \quad \operatorname{dn}\left\{u(1+k'), \frac{1-k'}{1+k'}\right\}.$$

4. Putting S for $\operatorname{sn} u \operatorname{dc} u$, prove that

$$\left(\frac{dS}{du}\right)^2 = 1 + 2(k'^2 - k^2)S^2 + S^4.$$

5. Verify that

$$\operatorname{sn}\left\{(1+k)u, \frac{2k^{\frac{1}{2}}}{1+k}\right\} = \frac{(1+k)s}{1+ks^2},$$

$$\operatorname{cn}\left\{(1+k)u, \frac{2k^{\frac{1}{2}}}{1+k}\right\} = \frac{cd}{1+ks^2},$$

$$\operatorname{dn}\left\{(1+k)u, \frac{2k^{\frac{1}{2}}}{1+k}\right\} = \frac{1-ks^2}{1+ks^2},$$

where s, c, d are $\operatorname{sn}(u, k), \operatorname{cn}(u, k), \operatorname{dn}(u, k)$, respectively.

6. If $k = \sqrt{2} - 1$, prove that

$$\operatorname{sn} u(-2)^{\frac{1}{2}} = (-2)^{\frac{1}{2}} \operatorname{sc} u \operatorname{nd} u,$$

$$\operatorname{cn} u(-2)^{\frac{1}{2}} = \operatorname{nc} u \operatorname{nd} u + k \operatorname{sc} u \operatorname{sd} u,$$

$$\operatorname{dn} u(-2)^{\frac{1}{2}} = \operatorname{nc} u \operatorname{nd} u - k \operatorname{sc} u \operatorname{sd} u.$$

Hence prove that for this value of k,

$$K'/K = \sqrt{2}.$$

7. If $k = \sin 75°$, verify that

$$\operatorname{sn} u(-3)^{\frac{1}{2}} = \iota \operatorname{sc} u(4\sqrt{3} - 6 - \operatorname{sn}^2 u)/(4 - 2\sqrt{3} - \operatorname{sn}^2 u),$$

$$\operatorname{cn} u(-3)^{\frac{1}{2}} = (2 - \sqrt{3})(2 - \sqrt{3}\operatorname{sn}^2 u)/\operatorname{cn} u(4 - 2\sqrt{3} - \operatorname{sn}^2 u),$$

$$\operatorname{dn} u(-3)^{\frac{1}{2}} = (2 - \sqrt{3})\operatorname{dc} u(2 - \operatorname{sn}^2 u)/(4 - 2\sqrt{3} - \operatorname{sn}^2 u).$$

Prove also that for this value of k,

$$K/K' = \sqrt{3}.$$

8. Find the expansions of $\operatorname{sn} u, \operatorname{cn} u, \operatorname{dn} u$ in ascending powers of u as far as u^5.

Ans. $\operatorname{sn} u = u - \tfrac{1}{6}(1+k^2)u^3 + \tfrac{1}{120}(1+14k^2+k^4)u^5\ldots,$
$\operatorname{cn} u = 1 - \tfrac{1}{2}u^2 + \tfrac{1}{24}(1+4k^2)u^4\ldots,$
$\operatorname{dn} u = 1 - \tfrac{1}{2}k^2 u^2 + \tfrac{1}{24}(4k^2+k^4)u^4\ldots.$

9. Trace the changes in sign and magnitude of sn, cn, dn for real and purely imaginary arguments for all real or purely imaginary values of k.

CHAPTER III.

ADDITION OF ARGUMENTS.

§ 33. We shall now show how to express the sn, cn, and dn of the sum of two arguments in terms of the elliptic functions of those arguments themselves.

Let u_1 and u_2 be the two arguments and let us write s_1, c_1, d_1 for sn u_1, cn u_1, dn u_1, and s_2, c_2, d_2 for sn u_2, cn u_2, dn u_2. This notation will often be found convenient.

Suppose u_1 and u_2 to vary in such a way that their sum is constant, say a.

Then $u_1 + u_2 = a$, $\dfrac{du_2}{du_1} = -1$.

Consider now some symmetric functions of u_1 and u_2, as sn u_1 + sn u_2, sn u_1 cn u_2 + sn u_2 cn u_1, etc.

We have

$$\frac{d}{du_1}(s_1+s_2) = c_1 d_1 - c_2 d_2,$$

$$\frac{d}{du_1}(s_1 c_2 + s_2 c_1) = c_1 d_1 c_2 - s_1 d_1 s_2 + s_1 s_2 d_2 - c_2 d_2 c_1$$
$$= (d_1 - d_2)(c_1 c_2 - s_1 s_2).$$

$$\frac{d}{du_1}(d_1 + d_2) = -k^2(s_1 c_1 - s_2 c_2)$$
$$= -k^2 s_1 c_1(c_2^2 + s_2^2) + k^2 s_2 c_2(c_1^2 + s_1^2)$$
$$= -k^2(c_1 c_2 - s_1 s_2)(s_1 c_2 - s_2 c_1).$$

Now $-k^2(s_1^2c_2^2-s_2^2c_1^2) = -k^2(s_1^2-s_2^2) = d_1^2-d_2^2$,
and thus we have

$$(d_1+d_2)\frac{d}{du_1}(s_1c_2+s_2c_1) = (s_1c_2+s_2c_1)\frac{d}{du_1}(d_1+d_2).$$

From this it follows at once that $\dfrac{s_1c_2+s_2c_1}{d_1+d_2} = $ a const. so long as $u_1+u_2=a$.

The value of this constant may be found by putting $u_1=0$ and $u_2=a$. It is $\dfrac{\operatorname{sn} a}{1+\operatorname{dn} a}$.

Thus
$$\frac{\operatorname{sn}(u_1+u_2)}{1+\operatorname{dn}(u_1+u_2)} = \frac{s_1c_2+s_2c_1}{d_1+d_2}.$$

§ 34. Again, $\dfrac{s_1c_2-s_2c_1}{d_1-d_2} = -\dfrac{1}{k^2}\cdot\dfrac{d_1+d_2}{s_1c_2+s_2c_1}$

$$= \text{a constant also}$$

$$= \frac{\operatorname{sn} a}{\operatorname{dn} a - 1}.$$

Thus $\quad -\dfrac{\operatorname{sn}(u_1+u_2)}{\operatorname{dn}(u_1+u_2)-1} = \dfrac{s_1c_2-s_2c_1}{d_1-d_2}.$

Inverting these two relations and subtracting, we have

$$\frac{2}{\operatorname{sn}(u_1+u_2)} = \frac{d_1+d_2}{s_1c_2+s_2c_1} - \frac{d_1-d_2}{s_1c_2-s_2c_1}$$

$$= \frac{2(s_1c_2d_2-s_2c_1d_1)}{s_1^2c_2^2-s_2^2c_1^2},$$

so that $\quad \operatorname{sn}(u_1+u_2) = \dfrac{s_1^2-s_2^2}{s_1c_2d_2-s_2c_1d_1}.$

ADDITION OF ARGUMENTS.

By inverting and adding, we have

$$\mathrm{ds}(u_1+u_2) = \frac{s_1 c_2 d_1 - s_2 c_1 d_2}{s_1^2 - s_2^2},$$

and

$$\mathrm{dn}(u_1+u_2) = \frac{s_1 c_2 d_1 - s_2 c_1 d_2}{s_1 c_2 d_2 - s_2 c_1 d_1}.$$

§ 35. In the same way we could prove the following relations

$$\frac{s_1 d_2 + s_2 d_1}{c_1 + c_2} = \frac{\mathrm{sn}(u_1+u_2)}{\mathrm{cn}(u_1+u_2)+1},$$

$$\frac{s_1 d_2 - s_2 d_1}{c_1 - c_2} = \frac{\mathrm{sn}(u_1+u_2)}{\mathrm{cn}(u_1+u_2)-1},$$

$$\frac{c_1 d_2 + c_2 d_1}{s_1 + s_2} = \frac{\mathrm{cn}(u_1+u_2) + \mathrm{dn}(u_1+u_2)}{\mathrm{sn}(u_1+u_2)},$$

$$\frac{c_1 d_2 - c_2 d_1}{s_1 - s_2} = \frac{\mathrm{cn}(u_1+u_2) - \mathrm{dn}(u_1+u_2)}{\mathrm{sn}(u_1+u_2)},$$

which we shall leave to the reader to verify.

§ 36. Any one of them is enough to give the value of $\mathrm{cn}(u_1+u_2)$. Adding the last two we have

$$\mathrm{cs}(u_1+u_2) = \frac{s_1 c_1 d_2 - s_2 c_2 d_1}{s_1^2 - s_2^2},$$

and hence

$$\mathrm{cn}(u_1+u_2) = \frac{s_1 c_1 d_2 - s_2 c_2 d_1}{s_1 c_2 d_2 - s_2 c_1 d_1},$$

by help of the value given in § 34 for $\mathrm{sn}(u_1+u_2)$.

§ 37. The formulae just found can be expressed in other ways.

We know that

$$\mathrm{sn}(u+\iota K') = \frac{1}{k}\,\mathrm{ns}\,u, \quad \mathrm{cn}(u+\iota K') = -\frac{\iota}{k}\,\mathrm{ds}\,u,$$

$$\mathrm{dn}(u+\iota K') = -\iota\,\mathrm{cs}\,u.$$

Put then $u_1 + \iota K'$ for u_1 in the above formulae. We have

$$\operatorname{sn}(u_1 + \iota K' + u_2) = \left(\frac{1}{k^2 s_1^2} - s_2^2\right) \Big/ \left(\frac{c_2 d_2}{k s_1} + s_2 \frac{c_1 d_1}{k s_1^2}\right),$$

$$\operatorname{sn}(u_1 + u_2) = \frac{1}{k \operatorname{sn}(u_1 + \iota K' + u_2)}$$

$$= \frac{s_1 c_2 d_2 + s_2 c_1 d_1}{1 - k^2 s_1^2 s_2^2},$$

$$\operatorname{cs}(u_1 + \iota K' + u_2) = \left(-\frac{\iota d_1 d_2}{k^2 s_1^2} + \frac{\iota s_2 c_2 c_1}{s_1}\right) \Big/ \left(\frac{1}{k^2 s_1^2} - s_2^2\right),$$

$$\operatorname{dn}(u_1 + u_2) = \iota \operatorname{cs}(u_1 + \iota K' + u_2)$$

$$= \frac{d_1 d_2 - k^2 s_1 s_2 c_1 c_2}{1 - k^2 s_1^2 s_2^2},$$

$$\operatorname{ds}(u_1 + \iota K' + u_2) = \left(-\frac{\iota c_1}{k s_1^2} c_2 + s_2 d_2 \cdot \frac{\iota d_1}{k s_1}\right) \Big/ \left(\frac{1}{k^2 s_1^2} - s_2^2\right).$$

The expression on the left is $-\iota k \operatorname{cn}(u_1 + u_2)$, so that

$$\operatorname{cn}(u_1 + u_2) = \frac{c_1 c_2 - s_1 s_2 d_1 d_2}{1 - k^2 s_1^2 s_2^2}.$$

These three forms, in which the denominator is $1 - k^2 s_1^2 s_2^2$, are those generally quoted. It may be verified by multiplication that they are the same as the former set. Thus, in the case of $\operatorname{dn}(u_1 + u_2)$,

$$(d_1 d_2 - k^2 s_1 s_2 c_1 c_2)(s_1 c_2 d_2 - s_2 c_1 d_1)$$
$$= s_1 c_2 d_1 (d_2^2 + k^2 s_2^2 c_1^2) - s_2 c_1 d_2 (d_1^2 + k^2 s_1^2 c_2^2)$$
$$= (s_1 c_2 d_1 - s_2 c_1 d_2)(1 - k^2 s_1^2 s_2^2),$$

for $\qquad d_2^2 + k^2 s_2^2 c_1^2 = 1 - k^2 s_1^2 s_2^2 = d_1^2 + k^2 s_1^2 c_2^2.$

The other verifications are left to the reader.

§ 38. By putting $u_1 + K$ for u_1 we may form another set from each of the two we have. The

ADDITION OF ARGUMENTS. 29

four sets of formulae are embodied in the following scheme:—

Numerator of $\operatorname{sn}(u_1+u_2)$:
$$s_1 c_2 d_2 + s_2 c_1 d_1, \quad s_1^2 - s_2^2, \quad s_1 c_1 d_2 + s_2 c_2 d_1, \quad s_1 c_2 d_1 + s_2 c_1 d_2.$$

Numerator of $\operatorname{cn}(u_1+u_2)$:
$$c_1 c_2 - s_1 s_2 d_1 d_2, \quad s_1 c_1 d_2 - s_2 c_2 d_1, \quad 1 - s_1^2 - s_2^2 + k^2 s_1^2 s_2^2, \quad c_1 c_2 d_1 d_2 - k'^2 s_1 s_2.$$

Numerator of $\operatorname{dn}(u_1+u_2)$:
$$d_1 d_2 - k^2 s_1 s_2 c_1 c_2, \quad s_1 c_2 d_1 - s_2 c_1 d_2, \quad c_1 c_2 d_1 d_2 + k'^2 s_1 s_2, \quad 1 - k^2 s_1^2 - k^2 s_2^2 + k^2 s_1^2 s_2^2.$$

Denominator of each:
$$1 - k^2 s_1^2 s_2^2, \quad s_1 c_2 d_2 - s_2 c_1 d_1, \quad c_1 c_2 + s_1 s_2 d_1 d_2, \quad d_1 d_2 + k^2 s_1 s_2 c_1 c_2.$$

§ 39. The above formulae give the sn, cn, dn of $u_1 - u_2$ by simply changing the sign of s_2.

Thus $\operatorname{sn}(u_1 - u_2) = \dfrac{s_1 c_2 d_2 - s_2 c_1 d_1}{1 - k^2 s_1^2 s_2^2}$, etc.

By combining different formulae we easily find the following, writing Δ for $1 - k^2 s_1^2 s_2^2$:—

$\Delta \operatorname{sn}(u_1+u_2)\operatorname{sn}(u_1-u_2) = s_1^2 - s_2^2,$

$\Delta \operatorname{cn}(u_1+u_2)\operatorname{cn}(u_1-u_2) = 1 - s_1^2 - s_2^2 + k^2 s_1^2 s_2^2,$

$\Delta \operatorname{dn}(u_1+u_2)\operatorname{dn}(u_1-u_2) = 1 - k^2 s_1^2 - k^2 s_2^2 + k^2 s_1^2 s_2^2,$

$\Delta \operatorname{sn}(u_1+u_2)\operatorname{cn}(u_1-u_2) = s_1 c_1 d_2 + s_2 c_2 d_1,$

$\Delta \operatorname{sn}(u_1+u_2)\operatorname{dn}(u_1-u_2) = s_1 c_2 d_1 + s_2 c_1 d_2,$

$\Delta \operatorname{cn}(u_1+u_2)\operatorname{dn}(u_1-u_2) = c_1 c_2 d_1 d_2 - k'^2 s_1 s_2.$

$\Delta\{1 \pm \operatorname{sn}(u_1+u_2)\}\{1 \pm \operatorname{sn}(u_1-u_2)\} = (c_2 \pm s_1 d_2)^2,$

$\Delta\{1 \pm k \operatorname{sn}(u_1+u_2)\}\{1 \pm k \operatorname{sn}(u_1-u_2)\} = (d_2 \pm k s_1 c_2)^2,$

$\Delta\{\operatorname{dn}(u_1+u_2) \pm \operatorname{cn}(u_1+u_2)\}\{\operatorname{dn}(u_1-u_2) \pm \operatorname{cn}(u_1-u_2)\}$
$$= (c_1 d_2 \pm c_2 d_1)^2,$$

$\Delta\{\operatorname{dn}(u_1+u_2) \pm k \operatorname{cn}(u_1+u_2)\}\{\operatorname{dn}(u_1-u_2) \pm k \operatorname{cn}(u_1-u_2)\}$
$$= (d_1 d_2 \pm k c_1 c_2)^2,$$

$\Delta\{1 \pm \operatorname{cn}(u_1+u_2)\}\{1 \pm \operatorname{cn}(u_1-u_2)\} = (c_1 \pm c_2)^2,$

$\Delta\{1 \pm \operatorname{dn}(u_1+u_2)\}\{1 \pm \operatorname{dn}(u_1-u_2)\} = (d_1 \pm d_2)^2,$

$\Delta\{1 \pm \operatorname{cn}(u_1+u_2)\}\{1 \mp \operatorname{cn}(u_1-u_2)\} = (s_1 d_2 \mp s_2 d_1)^2$

$$\Delta\{1 \pm \mathrm{dn}(u_1+u_2)\}\{1 \mp \mathrm{dn}(u_1-u_2)\} = k^2(s_1 c_2 \mp s_2 c_1)^2,$$
$$\Delta\{k' \pm \mathrm{dn}(u_1+u_2)\}\{k' \pm \mathrm{dn}(u_1-u_2)\} = (k' \pm d_1 d_2)^2,$$
$$\Delta\{k' \pm \mathrm{dn}(u_1+u_2)\}\{k' \mp \mathrm{dn}(u_1-u_2)\} = -k^2(c_1 c_2 \pm k' s_1 s_2)^2,$$
$$\Delta\{\mathrm{dn}(u_1+u_2) \pm k'\mathrm{sn}(u_1+u_2)\}\{\mathrm{dn}(u_1-u_2) \pm k'\mathrm{sn}(u_1-u_2)\}$$
$$= (c_2 d_1 \pm k' s_1)^2,$$
etc., etc.

The verification of the above results will give the reader useful practice in the algebraical handling of the elliptic functions.

§ 40. Since $u = v + a$ is the integral of the equation $du = dv$, a being the constant of integration, the different addition-formulae may be considered as forms of the integral of the same differential equation. Also if we write x for sn u, y for sn v, the differential equation becomes

$$(1-x^2)^{-\frac{1}{2}}(1-k^2 x^2)^{-\frac{1}{2}} dx = (1-y^2)^{-\frac{1}{2}}(1-k^2 y^2)^{-\frac{1}{2}} dy,$$

which therefore has an integral that is algebraical in x and y, although neither side can be integrated by means of algebraical functions. This fact was known for a long time before elliptic functions were invented. Euler succeeded in integrating the equation

$$X^{-\frac{1}{2}} dx + Y^{-\frac{1}{2}} dy = 0,$$

where X is a quartic function of x and Y is the same function of y.

Let
$$X = ax^4 + bx^3 + cx^2 + ex + f,$$
$$Y = ay^4 + by^3 + cy^2 + ey + f.$$

Then the integration is as follows:—
Write X', Y' for dX/dx and dY/dy.
We have
$$\frac{X-Y}{x-y} = a(x^3+x^2 y+xy^2+y^3)+b(x^2+xy+y^2)+c(x+y)+e.$$
$$X'+Y' = 4a(x^3+y^3)+3b(x^2+y^2)+2c(x+y)+2e.$$

EULER'S EQUATION.

Thus
$$\frac{X-Y}{x-y} - \tfrac{1}{2}(X'+Y') = -a(x+y)(x-y)^2 - \tfrac{1}{2}b(x-y)^2$$
$$= -(x-y)^2\{a(x+y)+\tfrac{1}{2}b\}.$$

Also $\quad \dfrac{d}{dx}X^{\frac{1}{2}} = \tfrac{1}{2}X^{-\frac{1}{2}}X', \quad \dfrac{d}{dy}Y^{\frac{1}{2}} = \tfrac{1}{2}Y^{-\frac{1}{2}}Y'.$

Hence
$$\frac{dx}{X^{\frac{1}{2}}} = \frac{dy}{-Y^{\frac{1}{2}}} = \frac{d(x+y)}{X^{\frac{1}{2}}-Y^{\frac{1}{2}}} = \frac{d(x-y)}{X^{\frac{1}{2}}+Y^{\frac{1}{2}}} = \frac{d(X^{\frac{1}{2}}-Y^{\frac{1}{2}})}{\tfrac{1}{2}(X'+Y')}$$
$$= \frac{(X^{\frac{1}{2}}-Y^{\frac{1}{2}})d(x-y) - (x-y)d(X^{\frac{1}{2}}-Y^{\frac{1}{2}})}{X-Y-\tfrac{1}{2}(X'+Y')(x-y)}$$
$$= \frac{1}{(x-y)\{a(x+y)+\tfrac{1}{2}b\}} d\left(\frac{X^{\frac{1}{2}}-Y^{\frac{1}{2}}}{x-y}\right).$$

Therefore
$$\{a(x+y)+\tfrac{1}{2}b\}d(x+y) = \frac{X^{\frac{1}{2}}-Y^{\frac{1}{2}}}{x-y} d\frac{X^{\frac{1}{2}}-Y^{\frac{1}{2}}}{x-y}$$

and
$$\left(\frac{X^{\frac{1}{2}}-Y^{\frac{1}{2}}}{x-y}\right)^2 = a(x+y)^2 + b(x+y) + g,$$

g being the constant of integration.

This is the integral sought.

Further information, with references, will be found in Forsyth's *Differential Equations*, pp. 237-247.

§ 41. Suppose in the addition-formulae that u_1 is real, and u_2 purely imaginary. Then s_1, c_1, d_1, c_2, d_2 are all real, and s_2 is purely imaginary. Thus the imaginary part of $\operatorname{sn}(u_1+u_2)$ is

$$\frac{s_2 c_1 d_1}{1-k^2 s_1{}^2 s_2{}^2}.$$

This cannot vanish unless $s_2=0$ or ∞, or $c_1=0$ or $d_1=0$.

But d_1 cannot vanish as u_1 is real, and if $c_1=0$ we have $u_1=$ an odd multiple of K.

Also since u_2 is purely imaginary, if $s_2 = 0$ or ∞ we have $u_2 = $ a multiple of $\iota K'$.

If then a complex argument have a real sn, its real part must be an odd multiple of K, or its imaginary part a multiple of $\iota K'$.

In the same way if the sn be purely imaginary, $s_1 = 0$ or ∞, or $c_2 = 0$ or $d_2 = 0$. These are all impossible but the first, so that the real part must be a multiple of $2K$.

§ 42. From this it follows that sn has no other period than $4K$ and $2\iota K'$. For if A were such a period it must be complex, say $A_1 + \iota A_2$. Then $\operatorname{sn}(u + A_1 + \iota A_2)$ is real or imaginary according as u is real or imaginary.

If u is real we have

$$A_2 = \text{a multiple of } K',$$

for $u + A_1$ is not generally an odd multiple of K.

If u is imaginary we have

$$A_1 = \text{a multiple of } 2K.$$

Hence there can be no periods other than those already found. The same holds for cn and dn.

§ 43. Suppose now that there are two arguments u_2 and u_3 for which sn, cn, and dn are all the same. Then it follows from the addition-formulae that

$$\operatorname{sn}(u_1 + u_2) = \operatorname{sn}(u_1 + u_3), \text{ etc.,}$$

whatever u_1 may be.

Hence $u_2 - u_3$ is a period for sn, cn, dn, and must be a quantity of the form $4mK + 4n\iota K'$.

Thus *all* arguments having the same sn as u are included in the formula

$$(-1)^m u + 2mK + 2n\iota K';$$

UNIFORMITY.

all having the same cn in the formula
$$\pm u + 4mK + 2n(K + \iota K');$$
and *all* having the same dn in the formula
$$\pm u + 2mK + 4n\iota K'.$$

§ 44. An important property of the elliptic functions, which has been assumed once or twice in the foregoing pages (as in § 41) is that they are *uniform*, that is to say that each of them has *one* single definite value for each value of its argument. Many examples might be given of functions for which this is not the case; $x^{\frac{1}{2}}$ is one.

The property may be proved as follows:—

Suppose sn $u = x$, and let us examine the behaviour of u and x when x is in the neighbourhood of a value a.

Put $x = a + \xi$, and let a be the value of u when $x = a$.

Then $\dfrac{du}{d\xi} = \{1 - (a+\xi)^2\}^{-\frac{1}{2}} \{1 - (ka + k\xi)^2\}^{-\frac{1}{2}}$.

The right hand side of this equation can be expanded in a series of powers of ξ, which will always converge absolutely so long as $|\xi|$ (the modulus of ξ) does not exceed the least of the quantities

$$\left| 1-a \right|, \quad \left| 1+a \right|, \quad \left| \frac{1}{k} - a \right|, \quad \left| \frac{1}{k} + a \right|.$$

(See Chrystal, *Algebra*, ch. xxvii., § 11).

By integrating every term on the right we get another absolutely convergent series since the term in ξ^r is multiplied by $\xi/(r+1)$, a constant (complex) multiple of a quantity that decreases as r increases.

Hence the value of u is given as the sum of an absolutely convergent series.

Therefore (see Chrystal, ch. xxx., § 18) ξ can be expanded in a convergent series of powers of $u - a$

within limits which are not infinitely narrow, and within those limits ξ is defined as a continuous uniform function of u (Chrystal, ch. xxvi., §§ 18, 19). This applies to every finite value of a but ± 1, $\pm 1/k$.

If a has any of these values we may put $x = a + \xi^2$, and deduce the same conclusion.

Lastly, in order to consider very great values of x we put $x = 1/\xi$, and find that $1/x$ is in that region a continuous uniform function of u.

Hence in all the plane there is no point where any branching-off of two or more values of x takes place, and therefore x is a uniform function of u.

The uniformity of $\operatorname{cn} u$ and $\operatorname{dn} u$ can be proved in the same way.

EXAMPLES ON CHAPTER III.

1. Verify from the formulae of this chapter that
$$\frac{d}{du_1} \operatorname{sn}(u_1 + u_2) = \operatorname{cn}(u_1 + u_2)\operatorname{dn}(u_1 + u_2),$$
$$\operatorname{cn}^2(u_1 + u_2) + \operatorname{sn}^2(u_1 + u_2) = 1,$$
$$\operatorname{dn}^2(u_1 + u_2) + k^2\operatorname{sn}^2(u_1 + u_2) = 1.$$

2. Find the sn, cn, dn of $u_1 + u_2 + u_3$ in terms of those of u_1, u_2, u_3, and show that the results are symmetrical.

3. If $u_1 + u_2 + u_3 = 0$, show that
$$d_1 d_2 d_3 - k^2 c_1 c_2 c_3 = k'^2,$$
$$d_1 + k^2 c_1 s_2 s_3 = d_2 d_3,$$
$$d_1 c_2 c_3 - c_1 d_2 d_3 = k'^2 s_2 s_3,$$
$$s_2 c_1 + s_3 d_1 + s_1 c_2 d_3 = 0,$$
$$c_1 + d_1 s_2 s_3 = c_2 c_3.$$

EXAMPLES III.

4. If $u_1+u_2+u_3+u_4=0$, show that
$$d_1d_2d_3d_4 - k^2 c_1c_2c_3c_4 + k^2k'^2 s_1s_2s_3s_4 = k'^2,$$

$$\begin{vmatrix} s_1 & c_1 & d_1 & 1 \\ s_2 & c_2 & d_2 & 1 \\ s_3 & c_3 & d_3 & 1 \\ s_4 & c_4 & d_4 & 1 \end{vmatrix} = 0,$$

$(s_1c_2-s_2c_1)(d_3-d_4)+(s_3c_4-s_4c_3)(d_1-d_2)=0,$
$(c_1d_2-c_2d_1)(s_3-s_4)+(c_3d_4-c_4d_3)(s_1-s_2)=0,$
$(s_1d_2-s_2d_1)(c_3-c_4)+(s_3d_4-s_4d_3)(c_1-c_2)=0.$

(These relations may be put in many more forms by such substitutions as u_1+K, u_2, u_3-K, u_4 for u_1, u_2, u_3, u_4.)

5. If $u_1+u_2+u_3=0$, then
$$\begin{vmatrix} s_1^3 & c_1d_1 & s_1 \\ s_2^3 & c_2d_2 & s_2 \\ s_3^3 & c_3d_3 & s_3 \end{vmatrix} = 0.$$

6. If $S(u)$ be written for sn u dc u and $S'(u)$ for its differential coefficient then
$$S(u_1+u_2) = \frac{S_1 S_2' + S_2 S_1'}{1 - S_1^2 S_2^2} = \frac{S_1^2 - S_2^2}{S_1 S_2' - S_2 S_1'}.$$

7. Verify the formulae of §39.

8. Prove the following:—
$$\text{sn } u \text{ sn } a = \frac{\text{cn}(u-a) - \text{cn}(u+a)}{\text{dn}(u-a) + \text{dn}(u+a)}$$
$$= \frac{1}{k^2} \cdot \frac{\text{dn}(u-a) - \text{dn}(u+a)}{\text{cn}(u-a) + \text{cn}(u+a)},$$
$$\text{cn } u \text{ cn } a = \frac{\text{cd}(u-a) + \text{cd}(u+a)}{\text{nd}(u-a) + \text{nd}(u+a)}$$
$$= -\frac{k'^2}{k^2} \cdot \frac{\text{nd}(u-a) - \text{nd}(u+a)}{\text{cd}(u-a) - \text{cd}(u+a)},$$

$$\operatorname{dn} u \operatorname{dn} a = \frac{\operatorname{dc}(u-a)+\operatorname{dc}(u+a)}{\operatorname{nc}(u-a)+\operatorname{nc}(u+a)}$$

$$= k'^2 \cdot \frac{\operatorname{nc}(u-a)-\operatorname{nc}(u+a)}{\operatorname{dc}(u-a)-\operatorname{dc}(u+a)},$$

$$\operatorname{sd} u \operatorname{cn} a = \frac{\operatorname{sn}(u+a)+\operatorname{sn}(u-a)}{\operatorname{dn}(u+a)+\operatorname{dn}(u-a)}$$

$$= \frac{1}{k^2} \cdot \frac{\operatorname{dn}(u-a)-\operatorname{dn}(u+a)}{\operatorname{sn}(u+a)-\operatorname{sn}(u-a)},$$

$$\operatorname{sc} u \operatorname{dn} a = \frac{\operatorname{sn}(u+a)+\operatorname{sn}(u-a)}{\operatorname{cn}(u+a)+\operatorname{cn}(u-a)}$$

$$= \frac{\operatorname{cn}(u-a)-\operatorname{cn}(u+a)}{\operatorname{sn}(u+a)-\operatorname{sn}(u-a)},$$

$$\operatorname{sn} u \operatorname{cd} a = \frac{\operatorname{sd}(u+a)+\operatorname{sd}(u-a)}{\operatorname{nd}(u+a)+\operatorname{nd}(u-a)}$$

$$= \frac{1}{k^2} \cdot \frac{\operatorname{nd}(u+a)-\operatorname{nd}(u-a)}{\operatorname{sd}(u+a)-\operatorname{sd}(u-a)},$$

$$\operatorname{sn} u \operatorname{dc} a = \frac{\operatorname{sc}(u+a)+\operatorname{sc}(u-a)}{\operatorname{nc}(u+a)+\operatorname{nc}(u-a)}$$

$$= \frac{\operatorname{nc}(u+a)-\operatorname{nc}(u-a)}{\operatorname{sc}(u+a)-\operatorname{sc}(u-a)},$$

$$\operatorname{dn} u \operatorname{nd} a = \frac{\operatorname{ds}(u+a)+\operatorname{ds}(u-a)}{\operatorname{ns}(u+a)+\operatorname{ns}(u-a)}$$

$$= \frac{\operatorname{ns}(u+a)-\operatorname{ns}(u-a)}{\operatorname{ds}(u+a)-\operatorname{ds}(u-a)},$$

$$\operatorname{sn} u \operatorname{ns} a = \frac{\operatorname{ds}(u-a)+\operatorname{ds}(u+a)}{\operatorname{cs}(u-a)-\operatorname{cs}(u+a)}$$

$$= \frac{\operatorname{cs}(u-a)+\operatorname{cs}(u+a)}{\operatorname{ds}(u-a)-\operatorname{ds}(u+a)},$$

EXAMPLES III.

$$\operatorname{sc} u \operatorname{nd} a = \frac{\operatorname{sd}(u+a)+\operatorname{sd}(u-a)}{\operatorname{cd}(u+a)+\operatorname{cd}(u-a)}$$

$$= -\frac{1}{k'^2} \cdot \frac{\operatorname{cd}(u+a)-\operatorname{cd}(u-a)}{\operatorname{sd}(u+a)-\operatorname{sd}(u-a)},$$

$$\operatorname{cn} u \operatorname{ds} a = \frac{\operatorname{dc}(u+a)+\operatorname{dc}(u-a)}{\operatorname{sc}(u+a)-\operatorname{sc}(u-a)}$$

$$= k'^2 \cdot \frac{\operatorname{sc}(u+a)+\operatorname{sc}(u-a)}{\operatorname{dc}(u+a)-\operatorname{dc}(u-a)},$$

$$\operatorname{cn} u \operatorname{nc} a = \frac{\operatorname{ns}(u-a)-\operatorname{ns}(u+a)}{\operatorname{cs}(u-a)-\operatorname{cs}(u+a)}$$

$$= \frac{\operatorname{cs}(u-a)+\operatorname{cs}(u+a)}{\operatorname{ns}(u-a)+\operatorname{ns}(u+a)}.$$

CHAPTER IV.

MULTIPLICATION AND DIVISION OF THE ARGUMENT.

§ 45. By putting $u_1 = u_2$ in the addition-formulae we easily find the values of $\operatorname{sn} 2u$, $\operatorname{cn} 2u$, $\operatorname{dn} 2u$ in terms of $\operatorname{sn} u$, $\operatorname{cn} u$, $\operatorname{dn} u$. Writing S, C, D, s, c, d for these quantities respectively, we have

$S = 2scd/(1 - k^2 s^4)$,
$C = (c^2 - s^2 d^2)/(1 - k^2 s^4) = (1 - 2s^2 + k^2 s^4)/(1 - k^2 s^4)$,
$D = (d^2 - k^2 s^2 c^2)/(1 - k^2 s^4) = (1 - 2k^2 s^2 + k^2 s^4)/(1 - k^2 s^4)$.

§ 46. Moreover, these equations can be solved for s, c, d if S, C, D are supposed known.

We have
$D - C = 2k'^2 s^2/(1 - k^2 s^4)$,
$D - k^2 C = k'^2(1 + k^2 s^4)/(1 - k^2 s^4)$,
$D - k^2 C + k'^2 = 2k'^2/(1 - k^2 s^4)$,
$D - k^2 C - k'^2 = 2k'^2 k^2 s^4/(1 - k^2 s^4)$.

Thus
$$s^2 = \frac{D - k^2 C - k'^2}{k^2(D - C)} = \frac{D - C}{D - k^2 C + k'^2}$$

$$= \frac{1 - C}{1 + D}, \text{ by subtraction,}$$

$$= \frac{1 - D}{k^2(1 + C)}, \text{ by subtracting again.}$$

HALVING OF THE ARGUMENT.

Hence we find the following formulae for $\tfrac{1}{2}u$:—

$$\operatorname{sn}\tfrac{1}{2}u = \left(\frac{1-\operatorname{cn}u}{1+\operatorname{dn}u}\right)^{\tfrac{1}{2}} = \frac{1}{k}\left(\frac{1-\operatorname{dn}u}{1+\operatorname{cn}u}\right)^{\tfrac{1}{2}}$$

$$= \frac{(1-\operatorname{cn}u)^{\tfrac{1}{2}}(1-\operatorname{dn}u)^{\tfrac{1}{2}}}{k\operatorname{sn}u},$$

$$\operatorname{cn}\tfrac{1}{2}u = \left(\frac{\operatorname{dn}u+\operatorname{cn}u}{1+\operatorname{dn}u}\right)^{\tfrac{1}{2}} = \frac{k'}{k}\left(\frac{1-\operatorname{dn}u}{\operatorname{dn}u-\operatorname{cn}u}\right)^{\tfrac{1}{2}}$$

$$= \frac{(1-\operatorname{dn}u)^{\tfrac{1}{2}}(\operatorname{dn}u+\operatorname{cn}u)^{\tfrac{1}{2}}}{k\operatorname{sn}u},$$

$$\operatorname{dn}\tfrac{1}{2}u = \left(\frac{\operatorname{dn}u+\operatorname{cn}u}{1+\operatorname{cn}u}\right)^{\tfrac{1}{2}} = k'\left(\frac{1-\operatorname{cn}u}{\operatorname{dn}u-\operatorname{cn}u}\right)^{\tfrac{1}{2}}$$

$$= \frac{(\operatorname{dn}u+\operatorname{cn}u)^{\tfrac{1}{2}}(1-\operatorname{cn}u)^{\tfrac{1}{2}}}{\operatorname{sn}u}.$$

§ 47. In particular

$$\operatorname{sn}\tfrac{1}{2}K = (1+k')^{-\tfrac{1}{2}},$$
$$\operatorname{cn}\tfrac{1}{2}K = k'^{\tfrac{1}{2}}(1+k')^{-\tfrac{1}{2}},$$
$$\operatorname{dn}\tfrac{1}{2}K = k'^{\tfrac{1}{2}},$$
$$\operatorname{sn}\tfrac{1}{2}\iota K' = \iota k^{-\tfrac{1}{2}},$$

being purely imaginary and of the same sign as its argument;

$$\operatorname{cn}\tfrac{1}{2}\iota K' = k^{-\tfrac{1}{2}}(1+k)^{\tfrac{1}{2}},$$

being a positive quantity;

$$\operatorname{dn}\tfrac{1}{2}\iota K' = (1+k)^{\tfrac{1}{2}},$$

being also positive.

These three may also be deduced from the others by using the complementary modulus.

ELLIPTIC FUNCTIONS.

Also

$$\operatorname{sn} \tfrac{1}{2}(K+\iota K') = \frac{1+\iota}{2} \frac{(1+k')^{\frac{1}{2}} - \iota(1-k')^{\frac{1}{2}}}{k^{\frac{1}{2}}},$$

$$\operatorname{cn} \tfrac{1}{2}(K+\iota K') = \frac{1-\iota}{2} \cdot \left(\frac{1+k}{kk'}\right)^{\frac{1}{2}} \{(1+k')^{\frac{1}{2}} - (1-k')^{\frac{1}{2}}\}$$

$$= (1-\iota)\left(\frac{k'}{2k}\right)^{\frac{1}{2}},$$

$$\operatorname{dn} \tfrac{1}{2}(K+\iota K') = \frac{1-\iota}{2} k'^{\frac{1}{2}} \{(1+k)^{\frac{1}{2}} + \iota(1-k)^{\frac{1}{2}}\}.$$

These three are most conveniently found from the former six by the addition-formulae.

MULTIPLICATION OF THE ARGUMENT BY ANY INTEGER.

§ 48. By repeated use of the addition-formulae we can find the elliptic functions of $3u, 4u, \ldots$, in terms of those of u.

We may prove the following facts about the formulae for sn nu, cn nu, dn nu :—

Firstly, when n is odd,

sn nu = sn u × a rational fractional function of $\operatorname{sn}^2 u$,
cn nu = cn u × a rational fractional function of $\operatorname{sn}^2 u$,
dn nu = dn u × a rational fractional function of $\operatorname{sn}^2 u$.

In each case the denominator is the same function, and is of the degree $n^2 - 1$ in sn u; the numerators are different, but are of the same degree, $n^2 - 1$.

Secondly, when n is even,

sn nu = sn u cn u dn u × a rational fractional function of $\operatorname{sn}^2 u$,

cn nu = a rational fractional function of $\operatorname{sn}^2 u$,
dn nu = a rational fractional function of $\operatorname{sn}^2 u$.

In each case the denominator is the same, and its degree is n^2 in $\operatorname{sn} u$; this is also the degree of the numerators of $\operatorname{cn} nu$ and $\operatorname{dn} nu$; the numerator of $\operatorname{sn} nu \div \operatorname{sn} u \operatorname{cn} u \operatorname{dn} u$ is of the degree $n^2 - 4$.

Clearly we may say a rational function of $\operatorname{cn}^2 u$ or $\operatorname{dn}^2 u$ instead of $\operatorname{sn}^2 u$ without altering the meaning or the degree to be assigned.

§ 49. These statements are evidently true when $n = 1$ or 2. Suppose them to be true for the values m and $m+1$ of n; one of these values will be even, and the other odd.

Write S_p, C_p, D_p, N_p for the three numerators and denominators of $\operatorname{sn} pu$, $\operatorname{cn} pu$, $\operatorname{dn} pu$ respectively, and s, c, d for $\operatorname{sn} u$, $\operatorname{cn} u$, $\operatorname{dn} u$. Then

$$S_{2m} = 2S_m C_m D_m N_m$$
$= scd \times$ a rational integral even function of s of degree $4m^2 - 4$,

$$C_{2m} = C_m^2 N_m^2 - S_m^2 D_m^2$$
$=$ a rational integral even function of s of degree $4m^2$,

$$D_{2m} = D_m^2 N_m^2 - k^2 S_m^2 C_m^2$$
$=$ a rational integral even function of s of degree $4m^2$,

$$N_{2m} = N_m^4 - k^2 S_m^4$$
$=$ a rational integral even function of s of degree $4m^2$.

Also
$$S_{2m+1} = S_m N_m C_{m+1} D_{m+1} + S_{m+1} N_{m+1} C_m D_m$$
$=$ a rational integral odd function of s of degree $2m^2 + 2(m+1)^2 - 1$, that is, $(2m+1)^2$;

$$C_{2m+1} = C_m N_m C_{m+1} N_{m+1} - S_m D_m S_{m+1} D_{m+1}$$
$=$ a similar function of c;

$$D_{2m+1} = D_m N_m D_{m+1} N_{m+1} - k^2 S_m C_m S_{m+1} C_{m+1}$$
$\qquad$ = a similar function of d;
$$N_{2m+1} = N_m^2 N_{m+1}^2 - k^2 S_m^2 S_{m+1}^2$$
$\qquad$ = a rational integral even function of s of degree $(2m+1)^2 - 1$.

Hence, if the theorems hold for the values $m, m+1$, they hold also for $2m$ and $2m+1$. Now they hold for 1 and 2, and therefore for 2 and 3, 4 and 5, and universally.

§ 50. Also these expressions will be in their lowest terms. Consider for instance C_m, a rational integral function of c of degree m^2. This must vanish whenever cn $mu = 0$, that is, whenever
$$mu = K + 2pK + 2q\iota K',$$
p and q being any integers.

Hence the roots of $C_m = 0$ as an equation for c are the values of cn$\dfrac{K + 2pK + 2q\iota K'}{m}$. This expression has m^2 different values found by making
$$p = 0, \quad 1 \ldots m-1,$$
and $\qquad q = 0, \pm 1 \ldots \pm \tfrac{1}{2}(m-1) \quad$ or $\quad \pm \tfrac{1}{2}m,$

in turn. Thus the degree of the numerator of cn mu cannot possibly be lower than m^2 and the expression we have found for cn mu is in its lowest terms.

Also as
$$C_m^2 + S_m^2 = N_m^2,$$
$$D_m^2 + k^2 S_m^2 = N_m^2,$$

and C_m, N_m have no common factor, S_m and D_m can have no factor in common with either.

§ 51. We may notice that when N_m is expressed in terms of s, the coefficient of s^2 in it vanishes.

For
$$N_{2m} = N_m^4 - k^2 S_m^4,$$
$$N_{2m+1} = N_m^2 N_{m+1}^2 - k^2 S_m^2 S_{m+1}^2.$$

Now s is a factor in S_m and S_{m+1}, so that if the term in s^2 is wanting in N_m and N_{m+1} it will be wanting in N_{2m} and N_{2m+1}.

Now $N_1 = 1$, $N_2 = 1 - k^2 s^4$, from which by induction the theorem follows.

By changing u into $u + \iota K'$ we find that the coefficient of s^{m^2-2} vanishes in S_m when m is odd and in N_m when m is even.

DIVISION OF THE ARGUMENT BY ANY INTEGER.

§ 52. If we know the value of sn u, the multiplication-formula gives us an equation to find sn u/n.

When n is odd,

sn $\dfrac{u}{n}$ is the root of an equation of the degree n^2, whose coefficients are rational in sn u.

When n is even,

$\operatorname{sn}^2 \dfrac{u}{n}$ is the root of a similar equation.

We may show that the solution of these equations depends only on that of equations of the nth degree.

§ 53. Take the case when n is odd.

Since sn $u = \operatorname{sn}(u + 4pK + 2q\iota K')$, it follows that sn $\dfrac{1}{n}(u + 4pK + 2q\iota K')$ is also a root, and as this expression has n^2 values it includes all the roots. Call it $\lambda(p, q)$.

Then clearly any symmetrical function of $\lambda(p, 0)$, $\lambda(p, 1), \ldots, \lambda(p, n-1)$ will be unchanged by adding any multiple of $2\iota K'$ to u. Such a function then will have only n values, given by putting $p = 0, 1, \ldots, n-1$

in turn. It will therefore be a root of an equation of the nth degree only.

Thus $\lambda(p, q)$ is the root of an equation of the nth degree whose coefficients are also given by equations of the nth degree, rational in sn u.

The same form of argument holds in the case when n is even, and also in the case when cn u or dn u is the function given and we have to find the sn, cn, or dn of u/n.

EXAMPLES ON CHAPTER IV.

1. Find the values of the sn, cn, and dn of $\frac{1}{2}(mK + niK')$ for all integral values of m and n.

2. Prove that sn $\frac{1}{3}K$ is a root of the equation
$$1 - 2x + 2k^2x^3 - k^2x^4 = 0.$$
What are the other roots, and which is the real one? *Ans.* $\operatorname{sn}(\frac{1}{3}K \pm \frac{2}{3}\iota K')$, $\operatorname{sn}(3K + \frac{2}{3}\iota K')$. The last is real.

3. With the notation of this chapter, show that $N_{2m+1} \pm C_{2m+1}$, expressed in terms of c, has $1 \pm c$ for a factor, the other factor being a perfect square.

4. Show that $N_{2m} - C_{2m}$ has $1 - c^2$ for a factor, and that the other factor of it is a perfect square, as is also $N_{2m} + C_{2m}$.

5. Prove that when expressed in terms of d, $N_{2m+1} \pm D_{2m+1}$ has $1 \pm d$ for a factor, the other factor being a perfect square, that $N_{2m} - D_{2m}$ has $1 - d^2$ for a factor, and that the other factor, as also $N_{2m} + D_{2m}$, is a perfect square.

6. Show that $N_{2m} \pm S_{2m}$ can be expressed as a perfect square, as can also the quotient of $N_{2m+1} \pm S_{2m+1}$ by $1 \pm (-1)^m s$.

7. Prove similar facts with regard to $N_m \pm kS_m$, $k'N_m \pm D_m$, $D_m \pm C_m$, $D_m \pm kC_m$.

EXAMPLES IV.

8. Prove that
$$(cN_m - C_m)^2 \div (N_{m+1} - C_{m+1})(N_{m-1} - C_{m-1}),$$
$$(dN_m - D_m)^2 \div (N_{m+1} - D_{m+1})(N_{m-1} - D_{m-1})$$
are independent of the argument u.

9. If μ, ν are any two nth roots of unity, show that the nth power of
$$\sum_{p=0}^{n-1} \sum_{q=0}^{n-1} \mu^p \nu^q \operatorname{sn} \frac{1}{n}(u + 4pK + 2q\iota K')$$
is a rational function of $\operatorname{sn} u$ and $\operatorname{cn} u \operatorname{dn} u$.

Hence show that the value of $\operatorname{sn} u/n$ may be found by the extraction of nth roots, if $\operatorname{sn} 2K/n$ and $\operatorname{sn} 2\iota K'/n$ are supposed known.

10. Use the last example to find expressions for
$$\operatorname{sn} \tfrac{1}{2}u, \quad \operatorname{sn} \tfrac{1}{3}u.$$

11. When n is odd, prove that
$$n \operatorname{sn} nu = \sum_{\nu=0}^{n-1} \sum_{\mu=0}^{n-1} \operatorname{sn}\left(u + \frac{4\mu K + 2\nu \iota K'}{n}\right)$$
and that $n^2 \operatorname{sn}^2 nu = \sum_{\nu=0}^{n-1} \sum_{\mu=0}^{n-1} \operatorname{sn}^2\left(u + \frac{4\mu K + 2\nu \iota K'}{n}\right).$

12. When n is even, prove that
$$n^2 \operatorname{ns}^2 nu = \sum_{\nu=0}^{n-1} \sum_{\mu=0}^{n-1} \operatorname{ns}^2\left(u + \frac{2\mu K + 2\nu \iota K'}{n}\right).$$

CHAPTER V.

INTEGRATION.

§ 54. We must now examine how far it is possible to integrate, with respect to u, any rational algebraic function of $\operatorname{sn} u$, $\operatorname{cn} u$, $\operatorname{dn} u$, or, as we shall write them, s, c, d.

In the first place, suppose the function to be $\dfrac{\phi(s,c,d)}{\psi(s,c,d)}$,

ϕ and ψ being rational integral algebraic functions.

We may make the denominator rational in s by multiplying it and the numerator by

$$\psi(s, -c, d)\psi(s, c, -d)\psi(s, -c, -d),$$

and by means of the relations

$$c^2 = 1 - s^2, \quad d^2 = 1 - k^2 s^2;$$

by means of the same relations we may reduce the numerator to the form

$$\chi_1(s) + c\chi_2(s) + d\chi_3(s) + cd\chi_4(s),$$

the denominator being $\chi(s)$ and χ, χ_1, χ_2, χ_3, χ_4, all rational integral algebraic functions.

§ 55. Now $\displaystyle\int \frac{cd\chi_4(s)}{\chi(s)} du = \int \frac{\chi_4(s)}{\chi(s)} ds,$

INTEGRATION.

which can be integrated by the ordinary rules for rational fractions;

$$\int \frac{d \cdot \chi_3(s)}{\chi(s)} du = \int \frac{\chi_3(s)}{\chi(s)} (1-s^2)^{-\frac{1}{2}} ds,$$

and this can be reduced to the integral of a rational function by the substitution

$$s = \frac{2z}{1+z^2},$$

which gives $\quad (1-s^2)^{\frac{1}{2}} = \dfrac{1-z^2}{1+z^2}.$

Also $\quad \displaystyle\int \frac{c\chi_2(s)}{\chi(s)} du = \int \frac{\chi_2(s)}{\chi(s)} (1-k^2 s^2)^{-\frac{1}{2}} ds,$

which can be reduced by putting

$$ks = \frac{2z}{1+z^2}.$$

The problem is thus reduced to the integration of $\chi_1(s)/\chi(s)$.

§ 56. The first step will naturally be the expression of $\chi_1(s)/\chi(s)$ as a series of partial fractions.

When this has been done the expressions to be integrated will fall under one of the two forms

$$s^m, \quad (s-a)^{-m},$$

a being any constant, real or imaginary. We will consider these in turn.

Let $\int s^m du = v_m.$ Now

$\dfrac{d}{du}(s^{m-3}cd)$

$= (m-3)s^{m-4}c^2 d^2 - s^{m-2} d^2 - k^2 s^{m-2} c^2$

$= (m-1)k^2 s^m - (m-2)(1+k^2)s^{m-2} + (m-3)s^{m-4},$

and therefore, integrating, we have

$$C + s^{m-3}cd$$
$$= (m-1)k^2 v_m - (m-2)(1+k^2)v_{m-2} + (m-3)v_{m-4}$$

where C is a constant.

Thus when $m > 3$, v_m can be expressed by means of v_{m-2} and v_{m-4}; and in the case when $m = 3$, v_3 can be expressed by means of v_1.

Thus when m is odd the integration of v_m depends only on that of v_1, and when m is even on that of v_2 and v_0.

§ 57. Now
$$v_1 = \int \operatorname{sn} u\, du$$
$$= 2 \int \operatorname{sn} 2x\, dx, \text{ putting } 2x = u,$$
$$= \int \frac{4 \operatorname{sn} x \operatorname{cn} x \operatorname{dn} x}{1 - k^2 \operatorname{sn}^4 x} dx$$
$$= 2 \int \frac{dz}{1 - k^2 z^2}, \text{ putting } z = \operatorname{sn}^2 x,$$
$$= \frac{1}{k} \log \frac{1 + kz}{1 - kz}$$
$$= \frac{1}{k} \log \frac{1 + k \operatorname{sn}^2 \tfrac{1}{2} u}{1 - k \operatorname{sn}^2 \tfrac{1}{2} u}.$$

Thus the integral of an odd power of sn u can always be expressed by means of the functions sn, cn, dn, log.

§ 58. Again,
$$v_0 = \int du = u,$$
$$v_2 = \int \operatorname{sn}^2 u\, du.$$

It is not possible to express v_2 by means of known functions, and a new symbol has to be introduced.

THE FUNCTION E.

The letter E is generally used, and the definition of its meaning is

$$Eu = \int_0^u \mathrm{dn}^2 u \, du,$$

so that $\qquad v_2 = (u - Eu)/k^2.$

The value of Eu when $u = K$ is generally denoted by E simply, so that

$$E = \int_0^K \mathrm{dn}^2 u \, du.$$

The Greek letter Z was used by Jacobi for a slightly different function, defined as follows:—

$$Zu = Eu - uE/K.$$

Thus $\qquad ZK = 0.$

One advantage in the use of this notation is that there is not the same risk of confusing the product Eu with the function Eu.

§ 59. We now turn to $\int (s-a)^{-m} du$, which we shall call w_m. Put $s - a = t$.

$$\frac{d}{du}(s-a)^{-m+1} cd$$

$$= (-m+1)(s-a)^{-m} c^2 d^2 - (s-a)^{-m+1} s(d^2 + k^2 c^2)$$

$$= t^{-m}[(-m+1) - (-m+1)(1+k^2)(t+a)^2$$

$$\quad + (-m+1)k^2(t+a)^4 - t(t+a)\{1 + k^2 - 2k^2(t+a)^2\}]$$

$$= -(m-1)(1-a^2)(1-k^2 a^2) t^{-m}$$

$$\quad + (2m-3)\{1 + k^2 - 2k^2 a^2\} a t^{-m+1}$$

$$\quad + (m-2)\{1 + k^2 - 6k^2 a^2\} t^{-m+2}$$

$$\quad - (2m-5) \cdot 2k^2 a \cdot t^{-m+3} - (m-3) \cdot k^2 \cdot t^{-m+4}.$$

Integrating, we find that w_m can be expressed by means of known functions, and w_{m-1}, w_{m-2}, w_{m-3},

w_{m-4}, provided always that $(m-1)(1-a^2)(1-k^2a^2)$ does not vanish.

If $a^2=1$ or $1/k^2$, then w_{m-1} can be expressed in terms of w_{m-2}, w_{m-3}, w_{m-4} for $2m-3$ does not vanish.

Hence for these special values of a the integral can be reduced to w_0, w_{-1}, w_{-2}, that is to v_0, v_1, v_2, and no new function need be introduced.

But in general the reduction can only be carried on as far as w_1, since when $m=1$ the coefficient of w_m in the formula of reduction vanishes. We must introduce a new function to express w_1, and w_2, w_3 ... can be expressed by means of this and known functions.

§ 60. Now though $\int (s-a)^{-1} du$ and $\int (s+a)^{-1} du$ cannot be found in terms of known functions, their sum can.

For by the addition-theorem
$$\operatorname{sn}(u+a) + \operatorname{sn}(u-a) = \frac{2 \operatorname{sn} u \operatorname{cn} a \operatorname{dn} a}{1 - k^2 \operatorname{sn}^2 a \operatorname{sn}^2 u}.$$

Now each of the terms on the left can be integrated since we have found $\int \operatorname{sn} u \, du$. Hence if a be so chosen that $k \operatorname{sn} a = 1/a$, we have an expression for
$$\int \frac{2s}{s^2 - a^2} du \quad \text{or} \quad \int (s-a)^{-1} du + \int (s+a)^{-1} du.$$

The new function that is introduced is therefore only needed to express
$$\int (s-a)^{-1} du - \int (s+a)^{-1} du,$$
and the one actually chosen is
$$\int_0^u \frac{k^2 \operatorname{sn} a \operatorname{cn} a \operatorname{dn} a \operatorname{sn}^2 u}{1 - k^2 \operatorname{sn}^2 u \operatorname{sn}^2 a} du.$$

THE FUNCTION Π.   51

This is denoted by $\Pi(u, a)$, and u is called the *argument*, a the *parameter*.

It has been shown then that any rational function of sn u, cn u, dn u can be integrated by help of the new functions E and Π. The properties of these will be considered in the next chapter.

EXAMPLES ON CHAPTER V.

1. Prove that $k^2 \int_{\frac{K}{2}}^{K} \text{sn}^2 u \, du = \int_{\frac{K}{2}}^{K} \text{ns}^2 u \, du - k'$.

2. Prove that $k^2 \int \dfrac{du}{1 - \text{sn}\, u} = \dfrac{\text{cn}\, u \, \text{dn}\, u}{1 - \text{sn}\, u} + k^2 u - Eu$.

3. Find $\int \dfrac{du}{1 + k \, \text{sn}\, u},\ \int \dfrac{du}{k' + \text{dn}\, u},\ \int \dfrac{du}{1 + \text{cn}\, u},\ \int \dfrac{du}{1 - \text{dn}\, u}$.

Ans. $\dfrac{1}{k'^2} Eu + \dfrac{k}{k'^2} \dfrac{\text{cn}\, u \, \text{dn}\, u}{1 + k \, \text{sn}\, u},\ \dfrac{1}{k^2 k'} Eu - \dfrac{k' u}{k^2} - \dfrac{\text{sn}\, u \, \text{cn}\, u}{k'(k' + \text{dn}\, u)}$,

$-\dfrac{\text{sn}\, u \, \text{dn}\, u}{1 + \text{cn}\, u} + u - Eu,\ \dfrac{1}{k^2}(u - Eu) - \dfrac{\text{sn}\, u \, \text{cn}\, u}{1 - \text{dn}\, u}$.

4. Show that

$$\int_0^u \dfrac{\text{sn}\, a \, \text{cn}\, a \, \text{dn}\, a \, du}{\text{sn}^2 u - \text{sn}^2 a} = \Pi(u, a) - \tfrac{1}{2} \log \dfrac{\text{sn}(a + u)}{\text{sn}(a - u)}.$$

5. Prove that

$$\int \text{ns}\, u \, du = \log \text{sn}\, \tfrac{1}{2} u - \log \text{cn}\, \tfrac{1}{2} u - \log \text{dn}\, \tfrac{1}{2} u,$$

$$\int \text{cs}\, u \, du = \log \text{sn}\, \tfrac{1}{2} u + \log \text{cn}\, \tfrac{1}{2} u - \log \text{dn}\, \tfrac{1}{2} u,$$

$$\int \text{ds}\, u \, du = \log \text{sn}\, \tfrac{1}{2} u - \log \text{cn}\, \tfrac{1}{2} u + \log \text{dn}\, \tfrac{1}{2} u.$$

6. Verify the formulae
$$\int_0^u \frac{\operatorname{sn} a \operatorname{cn} a \operatorname{dn} a \operatorname{dn}^2 u \, du}{\operatorname{cn}^2 u - \operatorname{sn}^2 a \operatorname{dn}^2 u} = \Pi(u, a) + \tfrac{1}{2} \log \frac{\operatorname{cn}(u-a)}{\operatorname{cn}(u+a)},$$
$$\int_0^u \frac{k^2 \operatorname{sn} a \operatorname{cn} a \operatorname{dn} a \operatorname{cn}^2 u \, du}{\operatorname{dn}^2 u - k^2 \operatorname{sn}^2 a \operatorname{cn}^2 u} = \Pi(u, a) + \tfrac{1}{2} \log \frac{\operatorname{dn}(u-a)}{\operatorname{dn}(u+a)}.$$

7. Prove that
$\Pi(ku, ka, 1/k) = \Pi(u, a, k),$
$\Pi(\iota u, \iota a, k') = \Pi(u, a) + \tfrac{1}{2} \log \dfrac{\operatorname{cn}(u-a)}{\operatorname{cn}(u+a)} - u \dfrac{\operatorname{sn} a \operatorname{dn} a}{\operatorname{cn} a},$
the modulus on the right being k throughout.

CHAPTER VI.

ADDITION OF ARGUMENTS FOR THE FUNCTIONS E, Π.

§ 61. Expressions can be found for $E(u_1+u_2)$ and $\Pi(u_1+u_2, a)$ in terms of functions of u_1 and u_2.

As in the former case, suppose $u_1+u_2=b$, a constant. Take the function Eu_1+Eu_2.

$$\frac{d}{du_1}(Eu_1+Eu_2) = d_1^2 - d_2^2$$
$$= -k^2(s_1^2 - s_2^2)$$
$$= -k^2\mathrm{sn}(u_1+u_2)(s_1c_2d_2 - s_2c_1d_1)$$
$$= k^2\mathrm{sn}\, b \cdot \frac{d}{du_1}(s_1 s_2).$$

Thus $Eu_1+Eu_2 - k^2 s_1 s_2 \mathrm{sn}\, b$ is constant, and putting $u_1=b$, $u_2=0$, we find its value to be Eb.

Hence

$$Eu_1 + Eu_2 - E(u_1+u_2) = k^2 \mathrm{sn}\, u_1 \mathrm{sn}\, u_2 \mathrm{sn}(u_1+u_2).$$

It follows that

$$Zu_1 + Zu_2 - Z(u_1+u_2) = k^2 \mathrm{sn}\, u_1 \mathrm{sn}\, u_2 \mathrm{sn}(u_1+u_2)..$$

§ 62. Putting $u_2 = K$ we have
$$E(u+K) - Eu = E - k^2 \mathrm{sn}\, u\, \mathrm{sn}(u+K)$$
$$= E - k^2 \mathrm{sn}\, u\, \mathrm{cd}\, u.$$

54 ELLIPTIC FUNCTIONS.

$$E(u+2K) - E(u+K) = E - k^2\operatorname{sn}(u+K)\operatorname{sn}(u+2K)$$
$$= E + k^2 \operatorname{sn} u \operatorname{cd} u.$$
$$E(u+2K) - Eu = 2E.$$

Hence $\quad E(u+2mK) - Eu = 2mE.$
$$Z(u+2mK) = Zu.$$

§ 63. Let us apply Jacobi's Imaginary Transformation (§ 21) to Eu.

We have

$$E(\iota u, k') = \iota \int_0^u \operatorname{dn}^2(\iota u, k') du = \iota \int_0^u \operatorname{dc}^2(u, k) du.$$

Now $\quad \dfrac{d}{du} \dfrac{\operatorname{sn} u \operatorname{dn} u}{\operatorname{cn} u} = \operatorname{dn}^2 u - k^2 \operatorname{sn}^2 u + \dfrac{\operatorname{sn}^2 u \operatorname{dn}^2 u}{\operatorname{cn}^2 u}$
$$= \operatorname{dc}^2 u - k^2 \operatorname{sn}^2 u.$$

Hence $\quad E(\iota u, k') = \iota \dfrac{\operatorname{sn} u \operatorname{dn} u}{\operatorname{cn} u} + \iota u - \iota Eu,$

the modulus when not expressed being k; no constant is added for both sides vanish with u.

Thus as $\operatorname{cn} K = 0$, $E(\iota K, k')$ and therefore also $E(\iota K', k)$ are infinite. Let us find the value of $E(K + \iota K', k)$.

$$E(K+u) = Eu + E - k^2 \operatorname{sn} u \operatorname{sn}(u+K)$$
$$= Eu + E - k^2 \operatorname{sn} u \operatorname{cd} u.$$

Thus
$$\iota E(K+u) + E(\iota u, k') = \iota(u+E) + \iota \operatorname{sn} u(\operatorname{dc} u - k^2 \operatorname{cd} u)$$
$$= \iota(u+E) + \dfrac{\iota k'^2 \operatorname{sn} u}{\operatorname{cn} u \operatorname{dn} u}.$$

Put now $\iota K'$ for u, and write
$$E' \text{ for } E(K', k').$$

Then $\quad \iota E(K + \iota K') - E' = \iota E - K',$
$$E(K + \iota K') = E + \iota(K' - E').$$

THE FUNCTION E.

§ 64. Since
$$E(K+u) = Eu + E - k^2 \operatorname{sn} u \operatorname{sn}(u+K)$$
we have
$$E(K+mK) = E(mK) + E = E(mK-K) + 2E = \ldots.$$
Thus $E(mK) = mE$ if m is any whole number. Also
$$E(u+2mK) - Eu = E(2mK) = 2mE.$$
In the same way
$$Em(K+\iota K') = mE(K+\iota K') = mE + \iota m(K'-E').$$
$$E(u+2mK+2m\iota K') - Eu = 2mE + 2m\iota(K'-E').$$
Thus
$$E(u+2mK+2n\iota K') = Eu + 2mE + 2n\iota(K'-E').$$

This equation shows that the effect on the function Eu of adding any multiple of $2K$ or $2\iota K'$ to its argument is to add the same multiple of $2E$ or $2\iota(K'-E')$ to the function.

§ 65. The quantities K, K', E, E' are connected by an important equation which we shall now prove.

Clearly
$$K\{E(K+\iota K') - E\} = \int_0^K \left(\int_K^{K+\iota K'} \operatorname{dn}^2 u \, du \right) dv,$$
$$\iota K' . E = \int_0^K \left(\int_K^{K+\iota K'} \operatorname{dn}^2 v \, du \right) dv,$$
Thus
$$K . E(K+\iota K') - (K+\iota K')E = \int_0^K \int_K^{K+\iota K'} (\operatorname{dn}^2 u - \operatorname{dn}^2 v) du \, dv.$$

The right-hand side may be transformed by putting
$$\operatorname{sn} u \operatorname{sn} v = x, \quad \operatorname{dn} u \operatorname{dn} v = y.$$

We have
$$\frac{\partial(x,y)}{\partial(u,v)} = \begin{vmatrix} \operatorname{cn} u\, \operatorname{dn} u\, \operatorname{sn} v, & -k^2 \operatorname{sn} u\, \operatorname{cn} u\, \operatorname{dn} v \\ \operatorname{cn} v\, \operatorname{dn} v\, \operatorname{sn} u, & -k^2 \operatorname{sn} v\, \operatorname{cn} v\, \operatorname{dn} u \end{vmatrix}$$
$$= -k^2 \operatorname{cn} u\, \operatorname{cn} v(\operatorname{sn}^2 v\, \operatorname{dn}^2 u - \operatorname{sn}^2 u\, \operatorname{dn}^2 v)$$
$$= \operatorname{cn} u\, \operatorname{cn} v(\operatorname{dn}^2 v - \operatorname{dn}^2 u).$$

The subject of integration is then $\dfrac{1}{\operatorname{cn} u\, \operatorname{cn} v}$.

Now $\quad k^2 \operatorname{cn}^2 u\, \operatorname{cn}^2 v - y^2 = k^2 k'^2 x^2 - k'^2,$

so that the transformed integral is

$$\iint \frac{k\, dy\, dx}{(y^2 + k^2 k'^2 x^2 - k'^2)^{\frac{1}{2}}}.$$

As to the limits, $\operatorname{sn} v$ takes all real values from 0 to 1, and $\operatorname{sn} u$ all real values from 1 to $1/k$.

Thus, if x has an assigned value >1, $\operatorname{sn} u$ and $\operatorname{sn} v$ are nearest when
$$\operatorname{sn} u = x, \quad \operatorname{sn} v = 1,$$
and furthest apart when
$$\operatorname{sn} u = 1/k, \quad \operatorname{sn} v = kx.$$

The value of y will therefore range from
$$k'(1 - k^2 x^2)^{\frac{1}{2}} \text{ to } 0.$$

For $\quad y^2 = 1 + k^4 x^2 - 2k^2 x - k^2(\operatorname{sn} u - \operatorname{sn} v)^2,$

which is least when $\operatorname{sn} u$ and $\operatorname{sn} v$ are furthest apart, and greatest when they are nearest.

Also, if x has an assigned value <1, $\operatorname{sn} u$ and $\operatorname{sn} v$ are nearest when
$$\operatorname{sn} u = 1, \quad \operatorname{sn} v = x,$$
and furthest apart when
$$\operatorname{sn} u = 1/k, \quad \operatorname{sn} v = kx.$$

VALUE OF AN INTEGRAL.

The value of y will therefore range from
$$k'(1-k^2x^2)^{\frac{1}{2}} \text{ to } 0 \text{ still.}$$

The integral is therefore
$$\int_0^{\frac{1}{k}}\int_0^{k'(1-k^2x^2)^{\frac{1}{2}}} \frac{k\,dy\,dx}{(y^2+k^2k'^2x^2-k'^2)^{\frac{1}{2}}}, \text{ that is, } -\frac{\iota\pi}{2}.$$

Any doubt there may be as to the sign of this result is removed by the consideration that in the original double integral
$$\mathrm{dn}\, v > k' > \mathrm{dn}\, u,$$
so that the subject of integration is always negative, while du is positive and dv has the sign $+\iota$.

Hence $K \cdot E(K+\iota K') - (K+\iota K')E = -\frac{1}{2}\iota\pi$.

Substituting the value that was found above for $E(K+\iota K')$, we have
$$EK' + E'K - KK' = \tfrac{1}{2}\pi.$$

§ 66. The following result will be useful afterwards:—
$$\int_0^K Eu\,du = \tfrac{1}{2}(KE - \log k').$$

We may prove it thus
$$\int_0^K Eu\,du = \int_0^K E(K-u)du = \tfrac{1}{2}\int_0^K \{Eu + E(K-u)\}du$$
$$= \tfrac{1}{2}\int_0^K \{E + k^2\mathrm{sn}\,u\,\mathrm{sn}\,K\,\mathrm{sn}(K-u)\}du$$
$$= \tfrac{1}{2}KE + \tfrac{1}{2}\int_0^K \frac{k^2\mathrm{sn}\,u\,\mathrm{cn}\,u}{\mathrm{dn}\,u}du$$
$$= \tfrac{1}{2}KE - \tfrac{1}{2}\log\mathrm{dn}\,K = \tfrac{1}{2}(KE - \log k').$$

ADDITION OF ARGUMENTS FOR THE FUNCTION Π.

§ 67. Again if $u_1 + u_2 = b$,

$$\frac{d}{du_1}\{\Pi(u_1, a) + \Pi(u_2, a)\}$$

$$= \frac{k^2\operatorname{sn} a \operatorname{cn} a \operatorname{dn} a s_1^2}{1 - k^2 s_1^2 \operatorname{sn}^2 a} - \frac{k^2\operatorname{sn} a \operatorname{cn} a \operatorname{dn} a s_2^2}{1 - k^2 s_2^2 \operatorname{sn}^2 a}$$

$$= \frac{k^2 \operatorname{sn} a \operatorname{cn} a \operatorname{dn} a (s_1^2 - s_2^2)}{(1 - k^2 s_1^2 \operatorname{sn}^2 a)(1 - k^2 s_2^2 \operatorname{sn}^2 a)}.$$

Now we have seen that

$$s_1^2 - s_2^2 = -\operatorname{sn} b \frac{d}{du_1}(s_1 s_2).$$

What we have to do is therefore to express $s_1^2 + s_2^2$ in terms of $s_1 s_2$ and b. Now

$$(1 - k^2 s_1^2 s_2^2)^2 \operatorname{cn} b \operatorname{dn} b$$
$$= (c_1 c_2 - s_1 s_2 d_1 d_2)(d_1 d_2 - k^2 s_1 s_2 c_1 c_2)$$
$$= c_1 c_2 d_1 d_2 (1 + k^2 s_1^2 s_2^2) - s_1 s_2 (k^2 c_1^2 c_2^2 + d_1^2 d_2^2),$$

$$(1 - k^2 s_1^2 s_2^2)^2 \operatorname{sn}^2 b$$
$$= 2 s_1 s_2 c_1 c_2 d_1 d_2 + s_1^2 c_2^2 d_2^2 + s_2^2 c_1^2 d_1^2.$$

So that

$$(1 - k^2 s_1^2 s_2^2)^2 \{(1 + k^2 s_1^2 s_2^2) \operatorname{sn}^2 b - 2 s_1 s_2 \operatorname{cn} b \operatorname{dn} b\}$$
$$= (1 + k^2 s_1^2 s_2^2)(s_1^2 c_2^2 d_2^2 + s_2^2 c_1^2 d_1^2) + 2 s_1^2 s_2^2 (k^2 c_1^2 c_2^2 + d_1^2 d_2^2)$$

which reduces to $(1 - k^2 s_1^2 s_2^2)^2 (s_1^2 + s_2^2)$.

Hence $s_1^2 + s_2^2 = (1 + k^2 s_1^2 s_2^2) \operatorname{sn}^2 b - 2 s_1 s_2 \operatorname{cn} b \operatorname{dn} b$, and

$$\frac{d}{du_1}\{\Pi(u_1, a) + \Pi(u_2, a)\}$$

$$= -\frac{k^2 \operatorname{sn} a \operatorname{cn} a \operatorname{dn} a \operatorname{sn} b}{1 - k^2 \operatorname{sn}^2 a \{(1 + k^2 s_1^2 s_2^2) \operatorname{sn}^2 b - 2 s_1 s_2 \operatorname{cn} b \operatorname{dn} b\} + k^4 s_1^2 s_2^2 \operatorname{sn}^4 a}$$

$$\times \frac{d}{du_1}(s_1 s_2).$$

THE FUNCTION Π.

The denominator
$$= (1 - k^2 \operatorname{sn}^2 a\, \operatorname{sn}^2 b) + 2k^2 s_1 s_2 \cdot \operatorname{sn}^2 a \operatorname{cn} b \operatorname{dn} b$$
$$+ k^4 s_1^2 s_2^2 \operatorname{sn}^2 a(\operatorname{sn}^2 a - \operatorname{sn}^2 b)$$
$$= (1 - k^2 \operatorname{sn}^2 a\, \operatorname{sn}^2 b)\{1 + k^2 s_1 s_2 \operatorname{sn} a \operatorname{sn}(a+b)\}$$
$$\{1 + k^2 s_1 s_2 \operatorname{sn} a \operatorname{sn}(a-b)\}.$$

The numerator
$$= \tfrac{1}{2}(1 - k^2 \operatorname{sn}^2 a\, \operatorname{sn}^2 b) k^2 \operatorname{sn} a \{\operatorname{sn}(a+b) - \operatorname{sn}(a-b)\}.$$

Hence
$$-\frac{d}{du_1}\{\Pi(u_1, a) + \Pi(u_2, a)\}$$
$$= \frac{1}{2} \frac{k^2 \operatorname{sn} a \operatorname{sn}(a+b)}{1 + k^2 s_1 s_2 \operatorname{sn} a \operatorname{sn}(a+b)} \frac{d}{du_1}(s_1 s_2)$$
$$- \frac{1}{2} \frac{k^2 \operatorname{sn} a \operatorname{sn}(a-b)}{1 + k^2 s_1 s_2 \operatorname{sn} a \operatorname{sn}(a-b)} \frac{d}{du_1}(s_1 s_2)$$
$$= \frac{1}{2} \frac{d}{du_1} \log \frac{1 + k^2 s_1 s_2 \operatorname{sn} a \operatorname{sn}(a+b)}{1 + k^2 s_1 s_2 \operatorname{sn} a \operatorname{sn}(a-b)}.$$

Integrating then, we have
$$\Pi(u_1 + u_2, a) - \Pi(u_1, a) - \Pi(u_2, a)$$
$$= \tfrac{1}{2} \log \frac{1 + k^2 \operatorname{sn} u_1 \operatorname{sn} u_2 \operatorname{sn} a \operatorname{sn}(u_1 + u_2 + a)}{1 - k^2 \operatorname{sn} u_1 \operatorname{sn} u_2 \operatorname{sn} a \operatorname{sn}(u_1 + u_2 - a)}.$$

§ 68. There is another interesting property of the function Π which we shall now prove. It connects $\Pi(u, a)$ with $\Pi(a, u)$, the same function with argument and parameter interchanged. We have

$$2 \frac{d}{du} \Pi(u, a) = \frac{2k^2 \operatorname{sn} a \operatorname{cn} a \operatorname{dn} a \operatorname{sn}^2 u}{1 - k^2 \operatorname{sn}^2 a \operatorname{sn}^2 u}$$
$$= k^2 \operatorname{sn} a \operatorname{sn} u \{\operatorname{sn}(u+a) + \operatorname{sn}(u-a)\}.$$

ELLIPTIC FUNCTIONS.

Thus

$$2\frac{\partial^2}{\partial u \partial a}\Pi(u, a)$$
$$= k^2 \operatorname{sn} u\{\operatorname{sn}(u+a)\operatorname{cn} a \operatorname{dn} a + \operatorname{sn} a \operatorname{cn}(u+a)\operatorname{dn}(u+a)\}$$
$$+ k^2 \operatorname{sn} u\{\operatorname{sn}(u-a)\operatorname{cn} a \operatorname{dn} a - \operatorname{sn} a \operatorname{cn}(u-a)\operatorname{dn}(u-a)\}.$$

But by the addition-theorem

$$\operatorname{sn} u = \frac{\operatorname{sn}^2(u+a) - \operatorname{sn}^2 a}{\operatorname{sn}(u+a)\operatorname{cn} a \operatorname{dn} a + \operatorname{sn} a \operatorname{cn}(u+a)\operatorname{dn}(u+a)}$$
$$= \frac{\operatorname{sn}^2(u-a) - \operatorname{sn}^2 a}{\operatorname{sn}(u-a)\operatorname{cn} a \operatorname{dn} a - \operatorname{sn} a \operatorname{cn}(u-a)\operatorname{dn}(u-a)},$$

for $u = (u+a) - a = (u-a) + a$.

Hence

$$2\frac{\partial^2}{\partial u \partial a}\Pi(u, a) = k^2 \operatorname{sn}^2(u+a) + k^2 \operatorname{sn}^2(u-a) - 2k^2 \operatorname{sn}^2 a$$
$$= 2 \operatorname{dn}^2 a - \operatorname{dn}^2(u-a) - \operatorname{dn}^2(u+a).$$

In the same way

$$2\frac{\partial^2}{\partial u \partial a}\Pi(a, u) = 2 \operatorname{dn}^2 u - \operatorname{dn}^2(a-u) - \operatorname{dn}^2(a+u),$$

so that $\frac{\partial^2}{\partial u \partial a}\{\Pi(u, a) - \Pi(a, u)\} = \operatorname{dn}^2 a - \operatorname{dn}^2 u$,

$$\frac{\partial}{\partial a}\{\Pi(u, a) - \Pi(a, u)\} = u \operatorname{dn}^2 a - Eu,$$

for $\Pi(0, a) = \Pi(a, 0) = 0$.

Finally then $\Pi(u, a) - \Pi(a, u) = u \cdot Ea - a \cdot Eu$.
This may also be written $uZa - aZu$.

EXAMPLES ON CHAPTER VI.

1. Prove that $E(u+K) - Eu = E + \frac{d}{du}\log \operatorname{dn} u$.

EXAMPLES. VI.

2. Prove that
$$E(u+K+\iota K') - Eu = E(K+\iota K') + \frac{d}{du}\log\operatorname{cn} u.$$

3. Prove that
$$E(u+\iota K') - Eu = E(K+\iota K') - E + \frac{d}{du}\log\operatorname{sn} u.$$

4. Prove that $kE(ku, 1/k) = E(u, k) - k'^2 u$.

5. Prove that
$$kE(\iota ku, \iota k'/k) = \iota u - \iota E(u, k) + \iota k^2 \operatorname{sn}(u, k)\operatorname{cd}(u, k).$$

6. Find the values of $E\tfrac{1}{2}K$, $E\tfrac{1}{2}\iota K'$, $E\tfrac{1}{2}(K+\iota K')$.

   Ans. $\tfrac{1}{2}(E+1-k')$, $\tfrac{1}{2}\iota(K'-E'+1+k)$,
   $\tfrac{1}{2}(E+\iota K'-\iota E'+k+\iota k')$.

7. Show that
$$\Pi(K, a) = KZa,$$
$$\Pi(K+\iota K', a) = (K+\iota K')Za + \iota\pi a/2K.$$

8. Prove the formula
$$2\Pi(u, a) = 2uEa - \int_{u-a}^{u+a} Ev\, dv.$$

9. Verify that
$$2\Pi(u, \tfrac{1}{2}K) = u(1-k') + \log\operatorname{dn}(u+\tfrac{1}{2}K) - \tfrac{1}{2}\log k'.$$

10. Prove that the limit when a is indefinitely diminished of $\Pi(u, a) \div a$ is $u - Eu$.

11. Show that $Enu - nEu$ is equal to a rational fractional function of $\operatorname{sn} u$ multiplied by $\operatorname{cn} u\, \operatorname{dn} u$.

By partial fractions or otherwise show that
$$nEnu - n^2 Eu = \frac{d}{du}\log N_n,$$
where N_n denotes the common denominator in the expressions for $\operatorname{sn} nu$, $\operatorname{cn} nu$, $\operatorname{dn} nu$.

62   ELLIPTIC FUNCTIONS.

12. In the same way prove the formula (n being odd)
$$nEnu - \sum_{\mu=0}^{n-1}\sum_{\nu=0}^{n-1} E\left(u + \frac{2\mu K + 2\nu\iota K'}{n}\right)$$
$$= -n(n-1)(E + \iota K' - \iota E').$$

13. Prove the formula for addition of parameters in the function Π, namely,
$$\Pi(u, a+b) - \Pi(u, a) - \Pi(u, b)$$
$$= \tfrac{1}{2}\log\frac{1 + k^2\operatorname{sn} a \operatorname{sn} b \operatorname{sn} u \operatorname{sn}(u+a+b)}{1 + k^2\operatorname{sn} a \operatorname{sn} b \operatorname{sn} u \operatorname{sn}(u-a-b)}$$
$$- k^2 u \operatorname{sn} a \operatorname{sn} b \operatorname{sn}(a+b).$$

14. Find the value of $\dfrac{\partial}{\partial a}\Pi(u, a)$ and prove that
$$\Pi(u, u) = uEu - \tfrac{1}{2}\int_0^{2u} Ev\, dv.$$

15. Prove, by putting $u + v = 2r$, $u - v = 2t$, and integrating, that
$$\Pi(u, a) + \Pi(v, a) - \Pi(u+v, a)$$
$$= \tfrac{1}{2}\log\frac{\{1 - k^2\operatorname{sn}^2(r-a)\operatorname{sn}^2 t\}\{1 - k^2\operatorname{sn}^2(r+a)\operatorname{sn}^2 r\}}{\{1 - k^2\operatorname{sn}^2(r+a)\operatorname{sn}^2 t\}\{1 - k^2\operatorname{sn}^2(r-a)\operatorname{sn}^2 r\}}.$$

CHAPTER VII.

WEIERSTRASS' NOTATION.

§ 69. For some purposes it is convenient to use the notation of Weierstrass, which we shall now explain shortly.

We write $\wp u$ for $a^2 \operatorname{ns}^2 au + \beta$, where a is any constant and β is a constant which we shall determine.

Differentiating, we have

$$\wp' u = -2a^3 \operatorname{ns} au \operatorname{cs} au \operatorname{ds} au.$$

Also
$$\operatorname{cs}^2 au = \operatorname{ns}^2 au - 1,$$
$$\operatorname{ds}^2 au = \operatorname{ns}^2 au - k^2.$$

Thus $(\wp' u)^2 = 4(\wp u - \beta)(\wp u - \beta - a^2)(\wp u - \beta - a^2 k^2).$

Now choose β so that the coefficient of $\wp^2 u$ on the right may vanish. Then

$$\beta = -\tfrac{1}{3} a^2 (1 + k^2),$$

and
$$(\wp' u)^2 = 4 \wp^3 u - g_2 \wp u - g_3,$$

where

$$g_2 = -4\beta(\beta + a^2) - 4\beta(\beta + a^2 k^2) - 4(\beta + a^2)(\beta + a^2 k^2),$$
$$g_3 = 4\beta(\beta + a^2)(\beta + a^2 k^2).$$

The equation
$$(\wp' u)^2 = 4 \wp^3 u - g_2 \wp u - g_3$$

with the particular equation
$$Lim_{u=0}(u^2 \wp u) = 1$$
constitutes the definition of Weierstrass' function $\wp u$.

§ 70. Conversely, if $\wp u = x$,
$$u = \int_x^\infty (4x^3 - g_2 x - g_3)^{-\frac{1}{2}} dx.$$

The periods of the function $\wp u$ are $2K/a$, $2\iota K'/a$. They are denoted by 2ω, $2\omega'$ respectively, and their sum by $2\omega''$. We then have
$$\wp\omega = \beta + a^2 = e_1, \text{ say},$$
$$\wp\omega'' = \beta + a^2 k^2 = e_2, \text{ say},$$
$$\wp\omega' = \beta = e_3, \text{ say};$$
and e_1, e_2, e_3 are the roots of the equation
$$4x^3 - g_2 x - g_3 = 0$$
in descending order of magnitude.

Thus $\qquad \wp'\omega = \wp'\omega' = \wp'\omega'' = 0.$

§ 71. We may write $\wp(u, g_2, g_3)$ for $\wp u$ when we wish to specify the quantities g_2, g_3.

Thus if we put μa for a in the original definitions $\wp u$ is changed into $\mu^2 \wp \mu u$, and g_2, g_3 are changed into $\mu^4 g_2$ and $\mu^6 g_3$.

Hence $\wp(u, g_2, g_3) = \mu^2 \wp(\mu u, \mu^{-4} g_2, \mu^{-6} g_3).$

In particular
$$\wp(\iota u, g_2, g_3) = -\wp(u, g_2, -g_3).$$

Also by a second differentiation we have
$$2\wp' u \wp'' u = 12 \wp^2 u \wp' u - g_2 \wp' u,$$
$$\wp'' u = 6\wp^2 u - \tfrac{1}{2} g_2.$$

WEIERSTRASS' NOTATION.

§ 72. The addition-formula for $\wp u$ is easily found from the formula
$$\operatorname{sn}(v_1+v_2) = \frac{s_1^2 - s_2^2}{s_1 c_2 d_2 - s_2 c_1 d_1}.$$

For
$$\operatorname{ns}^2(v_1+v_2) = \frac{(s_1 c_2 d_2 - s_2 c_1 d_1)^2}{(s_1^2 - s_2^2)^2},$$

$$\left(\frac{1}{s_1^2} - \frac{1}{s_2^2}\right)^2 \operatorname{ns}^2(v_1+v_2)$$

$$= \left(\frac{c_2 d_2}{s_1 s_2^2} - \frac{c_1 d_1}{s_1^2 s_2}\right)^2$$

$$= \left(\frac{c_2 d_2}{s_2^3} - \frac{c_1 d_1}{s_1^3}\right)^2 + \left(\frac{1}{s_1^2} - \frac{1}{s_2^2}\right)\left(\frac{c_2^2 d_2^2}{s_2^4} - \frac{c_1^2 d_1^2}{s_1^4}\right)$$

$$= \left(\frac{c_2 d_2}{s_2^3} - \frac{c_1 d_1}{s_1^3}\right)^2 - \left(\frac{1}{s_1^2} - \frac{1}{s_2^2}\right)^2 \left(\frac{1}{s_1^2} + \frac{1}{s_2^2} - 1 - k^2\right).$$

This, translated into Weierstrass' notation, as explained in § 69, gives, if we take $v_1 = au$, $v_2 = av$, and remember that $1 + k^2 = -3\beta/a^2$,

$$\wp(u+v) + \wp u + \wp v = \frac{1}{4}\left(\frac{\wp' u - \wp' v}{\wp u - \wp v}\right)^2,$$

the formula sought.

Again,
$$\frac{\partial}{\partial u} \frac{\wp' u - \wp' v}{\wp u - \wp v} = -\frac{\wp'^2 u - \wp' u \wp' v}{(\wp u - \wp v)^2} + \frac{6 \wp^2 u - \tfrac{1}{2} g_2}{\wp u - \wp v}$$

$$= -\frac{1}{2}\left(\frac{\wp' u - \wp' v}{\wp u - \wp v}\right)^2 - \frac{1}{2}\frac{\wp'^2 u - \wp'^2 v}{(\wp u - \wp v)^2} + \frac{6 \wp^2 u - \tfrac{1}{2} g_2}{\wp u - \wp v}.$$

Now $\wp'^2 u - \wp'^2 v = 4(\wp^3 u - \wp^3 v) - g_2(\wp u - \wp v),$

so that $\dfrac{\partial}{\partial u}\dfrac{\wp' u - \wp' v}{\wp u - \wp v} = 2(2\wp u + \wp v) - \dfrac{1}{2}\left(\dfrac{\wp' u - \wp' v}{\wp u - \wp v}\right)^2,$

and
$$\wp(u+v) = \wp u - \frac{1}{2}\frac{\partial}{\partial u}\frac{\wp' u - \wp' v}{\wp u - \wp v}.$$

D. E. F.

ELLIPTIC FUNCTIONS.

§ 73. Instead of the function E or Z, Weierstrass uses ζu, defined by the equation

$$\zeta u = \frac{1}{u} + \int_0^u \left(\frac{1}{u^2} - \wp u\right) du.$$

Differentiating, we find

$$\zeta' u = -\wp u.$$

The term $\frac{1}{u}$ is put outside the sign of integration because $\wp u$ is infinite at the lower limit, but $\wp u - \frac{1}{u^2}$ is finite.

The value of $\zeta(u+v)$ is found as follows:—

$$\zeta'(u+v) - \zeta' u = -\wp(u+v) + \wp u$$
$$= \frac{1}{2} \frac{\partial}{\partial u} \frac{\wp' u - \wp' v}{\wp u - \wp v}.$$

Hence $\quad \zeta(u+v) - \zeta u - C = \frac{1}{2} \frac{\wp' u - \wp' v}{\wp u - \wp v},$

where C is a quantity independent of u.

Also $\zeta u - \frac{1}{u} = 0$, when $u = 0$; and for the same value of u, $\wp' u + \frac{2}{u^3} = 0$, and $\wp u - \frac{1}{u^2}$ is finite.

Thus $\frac{1}{2} \frac{\wp' u - \wp' v}{\wp u - \wp v} + \frac{1}{u}$ is zero when $u = 0$ and

$$C = \zeta v.$$

Hence $\quad \zeta(u+v) - \zeta u - \zeta v = \frac{1}{2} \frac{\wp' u - \wp' v}{\wp u - \wp v}.$

The definition of ζ shows that since $\wp$ is an even function, ζ is an odd function. Thus

$$\zeta(-u) = -\zeta u;$$

and if $\quad u + v + w = 0,$

we have $\zeta u + \zeta v + \zeta w = -\dfrac{1}{2}\dfrac{\wp'u - \wp'v}{\wp u - \wp v}$
$= -\dfrac{1}{2}\dfrac{\wp'v - \wp'w}{\wp v - \wp w} = -\dfrac{1}{2}\dfrac{\wp'w - \wp'u}{\wp w - \wp u}$
$= -(\wp u + \wp v + \wp w)^{\frac{1}{2}}.$

The theory of these functions will be found developed in Halphen's *Traité des Fonctions Elliptiques et de leurs Applications* (Gauthier-Villars).

EXAMPLES ON CHAPTER VII.

1. Prove that
$\{\wp(u+\omega) - \wp\omega\}\{\wp u - \wp\omega\} = (\wp\omega - \wp\omega')(\wp\omega - \wp\omega'').$

2. If $u+v+w=0$, show that quantities a and b may be found such that
$$\wp'u = a\wp u + b,$$
$$\wp'v = a\wp v + b,$$
$$\wp'w = a\wp w + b.$$

3. In the last question prove that
$$a = -2(\zeta u + \zeta v + \zeta w).$$

4. If the equation $4x^3 - g_2 x - g_3 = 0$ has only one real root, prove that one corresponding value of k is a complex quantity whose modulus is unity, and that in this case $k^{\frac{1}{2}}\operatorname{sn} uk^{-\frac{1}{2}}$ is real if u is real.

5. Show that
$4\wp 2u = \wp u + \wp(u+\omega) + \wp(u+\omega') + \wp(u+\omega'').$

6. Prove the formulae

(1) $\{\wp u + \wp(u+\omega)\}\{\wp(u+\omega') + \wp(u+\omega'')\}$
$= -4\wp\omega\wp 2u - 4\wp\omega'\wp\omega''.$

(2) $\wp\tfrac{1}{2}u = \wp u + (\wp u - e_2)^{\frac{1}{2}}(\wp u - e_3)^{\frac{1}{2}}$
$+ (\wp u - e_3)^{\frac{1}{2}}(\wp u - e_1)^{\frac{1}{2}} + (\wp u - e_1)^{\frac{1}{2}}(\wp u - e_2)^{\frac{1}{2}}.$

ELLIPTIC FUNCTIONS.

7. Writing η, η', η'' for $\zeta\omega, \zeta\omega', \zeta\omega''$, prove the formulae

(1) $\eta + \eta' = \eta''$,

(2) $\zeta(u + 2m\omega + 2m'\omega') = \zeta u + 2m\eta + 2m'\eta'$,

if m and m' are integers;

(3) $\eta\omega' - \eta'\omega = \frac{1}{2}\iota\pi$,

(4) $\displaystyle\int_u^{u+\omega} \zeta v\, dv = (u-\omega')\eta + \int_{\omega'}^{\omega''} \zeta v\, dv + \frac{1}{2}\log\frac{e_1 - \wp u}{e_1 - e_3}$

$\qquad = (u + \frac{1}{2}\omega)\eta + \frac{1}{2}\log(\wp u - e_1)$
$\qquad\qquad - \frac{1}{4}\log(e_1 - e_2)(e_1 - e_3)$,

(5) $2\zeta\, 2u = \zeta u + \zeta(u+\omega) + \zeta(u+\omega') + \zeta(u-\omega'')$.

8. Show that
$$\int_0^u \frac{\wp' v}{\wp x - \wp v}\, dx - 2u\zeta v = \int_0^v \frac{\wp' u}{\wp x - \wp u}\, dx - 2v\, \zeta u.$$

9. If a and b have the same meaning as in Ex. 2, show that
$$\frac{d}{dx}\log\frac{\wp' x - a\,\wp x - b}{\wp' x + a\,\wp x + b} = \frac{\wp' u}{\wp x - \wp u} + \frac{\wp' v}{\wp x - \wp v} + \frac{\wp' w}{\wp x - \wp w} + a.$$

CHAPTER VIII.

DEGENERATION OF THE ELLIPTIC FUNCTIONS.

§ 74. For certain values of the modulus the elliptic functions degenerate into trigonometrical or exponential functions.

Thus let $k = 0$, then $\operatorname{dn} u = 1$ always, and

$$\frac{d}{du} \operatorname{sn} u = \operatorname{cn} u.$$

where $\operatorname{cn}^2 u + \operatorname{sn}^2 u = 1$,
and $\operatorname{sn} 0 = 0, \quad \operatorname{cn} 0 = 1.$

Therefore $\operatorname{sn} u$ is $\sin u$ and $\operatorname{cn} u$ is $\cos u$ (§ 6),

$$Eu = u, \quad K = E = \tfrac{1}{2}\pi, \quad Zu = 0.$$

§ 75. The six related moduli in this case are equal in pairs, the three values being $0, 1, \infty$.

If $k = 1$, then $\operatorname{dn} u = \operatorname{cn} u$, and we have

$$\frac{d}{du} \operatorname{sn} u = \operatorname{cn}^2 u = 1 - \operatorname{sn}^2 u, \quad \operatorname{sn} 0 = 0.$$

Put $\operatorname{sn} u = \tanh \theta$ and we have

$$\operatorname{sech}^2 \theta \frac{d\theta}{du} = 1 - \tanh^2 \theta = \operatorname{sech}^2 \theta.$$

Thus $\theta = u$, as they vanish together.

ELLIPTIC FUNCTIONS.

Hence $\text{sn}(u, 1) = \tanh u,$
$\text{cn}(u, 1) = \text{dn}(u, 1) = \text{sech } u,$

$$E(u, 1) = \int_0^u \text{dn}^2(u, 1) du = \text{sn}(u, 1) = \tanh u.$$

K is the least positive value of u for which sech $u = 0$, that is $K = \infty$,

$$E = \text{sn } K = 1.$$
$$Z(u, 1) = \tanh u.*$$

§ 76. For the case when $k = \infty$ we have

$$\text{sn}(u, k) = \frac{1}{k} \text{sn}\left(ku, \frac{1}{k}\right) = \frac{1}{k} \sin ku,$$

$$\text{cn}(u, k) = \text{dn}\left(ku, \frac{1}{k}\right) = 1,$$

$$\text{dn}(u, k) = \text{cn}\left(ku, \frac{1}{k}\right) = \cos ku.$$

These formulae show the behaviour of sn u, cn u, dn u when u is a quantity comparable with $1/k$.

The table of periods for the related moduli (§ 27) shows that in this case both the periods are infinite, their ratio being -1.

§ 77. When $k = 0$, the real quarter-period is finite, its value being $\frac{1}{2}\pi$; the imaginary period is infinite.

When $k = 1$, the imaginary quarter-period is finite and equal to $\frac{1}{2}\pi\iota$; the real period is infinite.

It may be shown that in this case the limit of $K \div \log k'$ is finite, and in fact $= -1$.

---

*The notation sg u, cg u for sn$(u, 1)$, cn$(u, 1)$ is sometimes used, in honour of Gudermann. As however the functions have names already, being the hyperbolic tangent and secant, we have not used the others.

The function arcsin tanh u is generally called the Gudermannian of u and written gd u. (See Chrystal's *Algebra*, chap. xxix., § 31, *note*.)

For we proved that
$$\tfrac{1}{2}(EK-\log k') = \int_0^K Eu\,du.$$

Thus $\quad \tfrac{1}{2}(EK-2K-\log k') = \int_0^K (Eu-1)du.$

Also $\quad E(u,\,1)-1 = \tanh u - 1$
$$= -2e^{-2u}/(1+e^{-2u}),$$
so that $\quad \int \{E(u,\,1)-1\}du = \log(1+e^{-2u}) = -\log 2,$
between the limits 0 and ∞. Hence
$$Lim_{k=1}(K/\log k') = Lim\,\frac{\log 4 - \log k'}{2-E} \div \log k' = -1,$$
as $E=1$ in the limit.

EXAMPLES ON CHAPTER VIII.

1. When k vanishes, prove that
$$\Pi(u,\,a+\iota K') = u\cot a + \tfrac{1}{2}\log\frac{\sin(a-u)}{\sin(a+u)}.$$

2. Show that
$\Pi(u,\,a,\,1) = \tfrac{1}{2}\log\cosh(u-a)\operatorname{sech}(u+a) + u\tanh a.$

3. Prove that the degeneration of $\wp u$ takes place when $g_2^3 = 27g_3^2$.

4. Show that $\operatorname{gd}(\iota\operatorname{gd} u) = \iota u$.

5. By the substitution
$$b\cot\theta - a\tan\theta = (a+b)\cot\phi,$$
prove that
$$\int_0^{\frac{\pi}{2}}(a^2\sin^2\theta + b^2\cos^2\theta)^{-\frac{1}{2}}d\theta = \int_0^{\frac{\pi}{2}}(a_1^2\sin^2\theta + b_1^2\cos^2\theta)^{-\frac{1}{2}}d\theta,$$
where $2a_1 = a+b$, $b_1 = a^{\frac{1}{2}}b^{\frac{1}{2}}$, and $a,\,b,\,a_1,\,b_1$ are all positive.

6. If in the last question a_2, b_2 are formed from a_1, b_1 as these from a, b, and if this process is carried on, show that in the limit, when n is increased indefinitely,

$$a_n = b_n = \frac{\pi}{2} \bigg/ \int_0^{\frac{\pi}{2}} (a^2\cos^2\theta + b^2\sin^2\theta)^{-\frac{1}{2}} d\theta.$$

(This quantity is Gauss' Arithmetico-Geometric Mean between a and b.)

CHAPTER IX.

DIFFERENTIATION WITH RESPECT TO THE MODULUS.

§ 78. The elliptic functions depend on two variables, the argument and the modulus. We must now show how to differentiate them with respect to the modulus. Write s, c, d for $\operatorname{sn} u$, $\operatorname{cn} u$, $\operatorname{dn} u$, and let σ, γ, δ denote

$$\frac{\partial}{\partial k}\operatorname{sn} u, \quad \frac{\partial}{\partial k}\operatorname{cn} u, \quad \frac{\partial}{\partial k}\operatorname{dn} u.$$

Since then
$$\frac{ds}{du} = cd,$$

we have
$$\frac{d\sigma}{du} = \gamma d + c\delta.$$

Since $\quad c^2 + s^2 = d^2 + k^2 s^2 = 1,$

we have $\quad c\gamma + s\sigma = 0, \quad d\delta + k s^2 + k^2 s \sigma = 0.$

Eliminating γ and δ,

$$cd\frac{d\sigma}{du} + s\sigma(d^2 + k^2 c^2) + k s^2 c^2 = 0.$$

Now
$$\frac{d}{du} cd = -s(d^2 + k^2 c^2),$$

so that
$$\frac{d}{du}\left(\frac{\sigma}{cd}\right) + \frac{k s^2}{d^2} = 0.$$

74 ELLIPTIC FUNCTIONS.

Again $\dfrac{d}{du}\dfrac{sc}{d} = c^2 - s^2 + \dfrac{k^2s^2c^2}{d^2} = c^2 - k'^2s^2/d^2$.

Thus $\dfrac{d}{du}\left(\dfrac{\sigma}{cd} - \dfrac{ksc}{k'^2d}\right) = -\dfrac{kc^2}{k'^2} = \dfrac{k'^2-d^2}{kk'^2}$,

and $\dfrac{\sigma}{cd} = \dfrac{ksc}{k'^2d} + \dfrac{u}{k} - \dfrac{Eu}{kk'^2}$,

each side vanishing when $u = 0$. Hence

$\dfrac{\partial \operatorname{sn} u}{\partial k} = \dfrac{k}{k'^2}\operatorname{sn} u \operatorname{cn}^2 u + \dfrac{u}{k}\operatorname{cn} u \operatorname{dn} u - \dfrac{Eu}{kk'^2}\operatorname{cn} u \operatorname{dn} u$,

$\dfrac{\partial \operatorname{cn} u}{\partial k} = -\dfrac{k}{k'^2}\operatorname{sn}^2 u \operatorname{cn} u - \dfrac{u}{k}\operatorname{sn} u \operatorname{dn} u + \dfrac{Eu}{kk'^2}\operatorname{sn} u \operatorname{dn} u$,

$\dfrac{\partial \operatorname{dn} u}{\partial k} = -\dfrac{k}{k'^2}\operatorname{sn}^2 u \operatorname{dn} u - ku \operatorname{sn} u \operatorname{cn} u + \dfrac{k}{k'^2}Eu \operatorname{sn} u \operatorname{cn} u$.

§ 79. From the last we may further find $\dfrac{\partial}{\partial k}Eu$, as follows:—

$\dfrac{\partial^2 Eu}{\partial u \partial k} = \dfrac{\partial}{\partial k}\operatorname{dn}^2 u = -\dfrac{2k}{k'^2}s^2d^2 - 2ku \cdot scd + \dfrac{2k}{k'^2}Eu \cdot scd$.

Now $\dfrac{d}{du}scd = c^2d^2 - s^2d^2 - k^2s^2c^2$

$= k'^2s^2 + c^2d^2 - 2s^2d^2$,

$\dfrac{d}{du}(Eu \cdot s^2) = 2Eu \cdot scd + s^2d^2$,

$\dfrac{d}{du}u \cdot s^2 = 2u \cdot scd + s^2$.

Hence $\dfrac{\partial^2 Eu}{\partial u \partial k} - \dfrac{d}{du}\left\{\dfrac{ks^2}{k'^2}Eu - ks^2 \cdot u + \dfrac{kscd}{k'^2}\right\}$

$= -\dfrac{k}{k'^2}s^2d^2 + ks^2 - \dfrac{k}{k'^2}(k'^2s^2 + c^2d^2) = -\dfrac{k}{k'^2}d^2$.

DIFFERENTIATION OF THE PERIODS. 75

Integrating,
$$\frac{\partial}{\partial k}Eu = -\frac{k}{k'^2}\operatorname{cn}^2 u\, Eu - ku\operatorname{sn}^2 u + \frac{k}{k'^2}\operatorname{sn} u \operatorname{cn} u \operatorname{dn} u,$$
since again both sides vanish with u.

§ 80. These equations enable us also to find
$$\frac{dK}{dk}, \quad \frac{dE}{dk}, \quad \text{etc.}$$

We have $\operatorname{cn}(K, k) = 0$, and therefore
$$-\operatorname{sn} K \operatorname{dn} K \cdot \frac{dK}{dk} - \frac{K}{k}\operatorname{sn} K \operatorname{dn} K + \frac{E}{kk'^2}\operatorname{sn} K \operatorname{dn} K = 0$$
by differentiating. Thus $\dfrac{dK}{dk} = \dfrac{E - k'^2 K}{kk'^2}$.

Again, when $u = K$, $\quad \dfrac{\partial}{\partial k}Eu = -kK$.

Thus
$$\frac{dE}{dk} = -kK + \operatorname{dn}^2 K \frac{dK}{dk}$$
$$= -kK + \frac{E - k'^2 K}{k} = \frac{E - K}{k}.$$

Also $\quad \dfrac{dK'}{dk'} = \dfrac{E' - k^2 K'}{k' k^2}, \quad \dfrac{dE'}{dk'} = \dfrac{E' - K'}{k'};$

so that $\quad \dfrac{dK'}{dk} = \dfrac{k^2 K' - E'}{kk'^2}, \quad \dfrac{dE'}{dk} = \dfrac{k(K' - E')}{k'^2}.$

§ 81. Again
$$\frac{d}{dk}\left(kk'^2 \frac{dK}{dk}\right) = \frac{d}{dk}(E - k^2 K)$$
$$= \frac{E - K}{k} - k^2 \cdot \frac{E - k'^2 K}{kk'^2} + 2kK$$
$$= kK.$$

Putting $k^2 = c$, $k'^2 = c'$ in this we have it in the form

$$\frac{d}{dc}\left(cc'\frac{dK}{dc}\right) = \tfrac{1}{4}K,$$

which is unchanged if c and c' are interchanged. It must therefore also hold when K' is put for K, as can easily be verified.

The most general solution of the equation

$$\frac{d}{dk}\left\{(k-k^3)\frac{dy}{dk}\right\} = ky$$

is accordingly $\qquad y = AK + BK',$

where A and B are any constants.

§ 82. In the same way

$$\frac{d}{dk}\left(k\frac{dE}{dk}\right) = \frac{dE}{dk} - \frac{dK}{dk} = \frac{E-K}{k} - \frac{E-k^2K}{kk'^2} = -\frac{kE}{k'^2}.$$

This equation is not satisfied by E' also, but we saw (§ 63) that

$$E(K + \iota K') = E + \iota(K' - E'),$$

so that $K' - E'$ is suggested as a second solution.

Now $\qquad \dfrac{d}{dk}(K' - E') = -\dfrac{E'}{k}.$

Thus $\dfrac{d}{dk}\left\{k\dfrac{d}{dk}(K' - E')\right\} = -\dfrac{dE'}{dk} = -\dfrac{k}{k'^2}(K' - E').$

Hence the most general solution of the equation

$$(1-k^2)\frac{d}{dk}\left(k\frac{dz}{dk}\right) + kz = 0$$

is $\qquad z = CE + D(K' - E'),$

C and D being any constants.

EXPANSION OF THE PERIODS.

§ 83. The differential equations just found for K and E may be solved in series, and thus the expansions of K, K', E, E' in powers of k may be found.

Take
$$\frac{d}{dk}\left\{(k-k^3)\frac{dy}{dk}\right\} = ky,$$

and put
$$y = \sum_{r=0}^{r=\infty} \mu_r k^{s+2r},$$

for the exponents of k in successive terms must clearly differ by 2. Then

$$\frac{d}{dk}\left\{(k-k^3)\frac{dy}{dk}\right\} - ky = s^2 \mu_0 k^{s-1}$$
$$+ \Sigma\{(s+2r)^2 \mu_r - (s+2r-2)(s+2r)\mu_{r-1} - \mu_{r-1}\} k^{s+2r-1}.$$

The coefficients are therefore given successively by the relation $\mu_r = \left(\dfrac{s+2r-1}{s+2r}\right)^2 \mu_{r-1}$, and the values of s by the equation $s^2 = 0$. This equation has equal roots, so that we find the second solution by differentiating the first, namely

$$k^s + \sum_{r=1}^{\infty} \frac{(s+1)^2(s+3)^2 \ldots (s+2r-1)^2}{(s+2)^2(s+4)^2 \ldots (s+2r)^2} k^{s+2r},$$

with respect to s before putting in the value of s.

Hence, if

$$y_1 = 1 + \left(\frac{1}{2}\right)^2 k^2 + \left(\frac{1.3}{2.4}\right)^2 k^4 + \ldots + \left\{\frac{1.3\ldots(2r-1)}{2.4\ldots 2r}\right\}^2 k^{2r} + \ldots,$$

and $\quad y_2 = y_1 \log k + 2 \sum_{r=1}^{\infty} \left\{\dfrac{1.3\ldots(2r-1)}{2.4\ldots 2r}\right\}^2$
$$\times \left(1 - \frac{1}{2} + \frac{1}{3} - \frac{1}{4} + \ldots - \frac{1}{2r}\right) k^{2r},$$

the complete primitive is
$$y = A y_1 + B y_2.$$

ELLIPTIC FUNCTIONS.

§ 84. We may therefore choose A and B so that this expression shall be the value of K or K'.

Now we have seen that when $k=0$, $K=\tfrac{1}{2}\pi$. But $y_1=1$, $y_2=\infty$ for this value of k. Thus

$$K = \tfrac{1}{2}\pi y_1 = \frac{\pi}{2}\left\{1 + \left(\frac{1}{2}\right)^2 k^2 + \left(\frac{1\cdot 3}{2\cdot 4}\right)^2 k^4 + \ldots\right\}.$$

Suppose that $\quad K' = A y_1 + B y_2.$

§ 85. In the same way, from the equation for E we may find series for E and $K'-E'$, or we may use the formulae

$$E = k'^2 K + kk'^2 dK/dk,$$
$$K' - E' = k'^2 K' + kk'^2 dK'/dk.$$

Putting
$$z_1 = (1-k^2)\left(y_1 + k\frac{dy_1}{dk}\right),$$
$$z_2 = (1-k^2)\left(y_2 + k\frac{dy_2}{dk}\right),$$

we find
$$E = \tfrac{1}{2}\pi z_1,$$
$$K' - E' = A z_1 + B z_2,$$

where
$$z_1 = 1 - \frac{1}{2^2}k^2 - \frac{1^2\cdot 3}{2^2\cdot 4^2}k^4 - \ldots - \frac{1^2\cdot 3^2\ldots(2r-3)^2(2r-1)}{2^2\cdot 4^2\ldots(2r)^2}k^{2r} - \ldots,$$

$$z_2 = z_1 \log k + 1 + \sum_{r=1}^{r=\infty} \frac{k^{2r}}{2r-1}\left\{\frac{1\cdot 2\ldots(2r-1)}{2\cdot 4\ldots 2r}\right\}^2$$
$$\times\left[\frac{1}{2r-1} - 2\left(1 - \frac{1}{2} + \ldots - \frac{1}{2r}\right)\right].$$

Hence $\quad E' = A(y_1 - z_1) + B(y_2 - z_2);$
and as when $k = 0$,
$$E' = 1, \quad y_1 - z_1 = 0, \quad \text{and} \quad y_2 - z_2 = -1,$$
we have $\quad B = -1.$

EXAMPLES IX.

§ 86. A, as well as B, may be found as follows:—
We found (§ 66) that the limit of $\frac{1}{2}(EK - 2K - \log k')$, when $k=1$, was $-\log 2$.
Thus in the limit, when $k=0$,

$$\{A(y_1 - z_1) + B(y_2 - z_2)\}(Ay_1 + By_2)$$
$$- 2(Ay_1 + By_2) - \log k + 2\log 2 = 0.$$

The coefficient of $\log k$ on the left is $-B^2 - 2B - 1$. This must vanish, so that, as we found before,

$$B = -1.$$

The absolute term is $-AB - 2A + 2\log 2$. This must vanish, so that

$$A = 2\log 2.$$

Hence
$$K'' = 2y_1 \log 2 - y_2,$$
$$E'' = 2(y_1 - z_1)\log 2 - (y_2 - z_2).$$

It is noticeable that the series y_1, z_1 are hypergeometric. Thus, in the notation of hypergeometric series,

$$K = \frac{\pi}{2} F(\tfrac{1}{2}, \tfrac{1}{2}, 1, k^2),$$

$$E = \frac{\pi}{2} F(-\tfrac{1}{2}, \tfrac{1}{2}, 1, k^2).$$

EXAMPLES ON CHAPTER IX.

1. Prove that K increases with k so long as the latter is a positive proper fraction, while E decreases as k increases.

2. Show that

$$\frac{\partial}{\partial k}\int_0^u Ev\, dv = -\frac{1}{2kk'^2}(Eu)^2 + \frac{u}{k}Eu - \frac{u^2}{2k} + \frac{k\,\mathrm{sn}^2 u}{2k'^2},$$

and hence find $\dfrac{\partial}{\partial k}\Pi(u, a)$.

80   ELLIPTIC FUNCTIONS.

3. Prove that if N_n is the common denominator of sn nu, cn nu, dn nu, and is equal to unity when $u=0$, then

$$2n^2kk'^2\frac{\partial N_n}{\partial k} + \frac{\partial^2 N_n}{\partial u^2} + 2n^2\frac{\partial N_n}{\partial u}(Eu - k'^2 u)$$
$$+ n^2(n^2-1)N_n k^2 \operatorname{sn}^2 u = 0.$$

4. Writing x for sn u, transform this differential equation into the following, in which x and k are the independent variables:—

$$2n^2 kk'^2 \frac{\partial N_n}{\partial k} + (1-x^2)(1-k^2 x^2)\frac{\partial^2 N_n}{\partial x^2}$$
$$+ x\frac{\partial N_n}{\partial x}\{(2n^2-1)k^2(1-x^2)-1+k^2 x^2\} + n^2(n^2-1)N_n k^2 x^2 = 0.$$

(For Examples 3 and 4 use the result of Ex. 11, Chap. VI.)

5. Show that

$$(g_2^3 - 27g_3^2)\frac{\partial}{\partial g_2}\wp u$$
$$= \wp'u(\tfrac{1}{4}g_2^2 u - \tfrac{9}{2}g_3\zeta u) - 9g_3\wp^2 u + \tfrac{1}{2}g_2^2\wp u + \tfrac{3}{2}g_2 g_3,$$

$$(g_2^3 - 27g_3^2)\frac{\partial}{\partial g_3}\wp u$$
$$= \wp'u(3g_2\zeta u - \tfrac{9}{2}g_3 u) + 6g_2\wp^2 u - 9g_3\wp u - g_2^2.$$

6. Prove also that

$$(g_2^3 - 27g_3^2)\frac{\partial}{\partial g_2}\zeta u$$
$$= -\wp u(\tfrac{1}{4}g_2^2 u - \tfrac{9}{2}g_3\zeta u) + \tfrac{9}{4}g_3\wp'u + \tfrac{1}{4}g_2^2 \zeta u - \tfrac{3}{8}g_2 g_3 u,$$

and that

$$(g_2^3 - 27g_3^2)\frac{\partial}{\partial g_3}\zeta u$$
$$= -\wp u(3g_2\zeta u - \tfrac{9}{2}g_3 u) - \tfrac{3}{2}g_2\wp'u - \tfrac{9}{2}g_3\zeta u + \tfrac{1}{4}g_2^2 u.$$

EXAMPLES IX.

7. Show, by differentiating the equation
$$\wp(u+2\omega)=\wp u,$$
or otherwise, that
$$(g_2{}^3-27g_3{}^2)\frac{\partial \omega}{\partial g_2} = -\tfrac{1}{4}\omega g_2{}^2+\tfrac{9}{2}\eta g_3,$$

$$(g_2{}^3-27g_3{}^2)\frac{\partial \omega}{\partial g_3} = \tfrac{9}{2}\omega g_3 - 3\eta g_2.$$

8. Prove also that
$$(g_2{}^3-27g_3{}^2)\frac{\partial \eta}{\partial g_2} = \tfrac{1}{4}\eta g_2{}^2-\tfrac{3}{8}\omega g_2 g_3,$$

$$(g_2{}^3-27g_3{}^2)\frac{\partial \eta}{\partial g_3} = -\tfrac{9}{2}\eta g_3 + \tfrac{1}{4}\omega g_2{}^2.$$

9. Verify by differentiating that $EK'+E'K-KK'$ and $\eta\omega'-\eta'\omega$ are constants.

10. Interpret the following differential equation, satisfied by $\wp u$:—
$$2g_2\frac{\partial}{\partial g_2}\wp u+3g_3\frac{\partial}{\partial g_3}\wp u = \tfrac{1}{2}u\wp'u+\wp u.$$

11. Verify the values of $\dfrac{dK}{dk}$ and $\dfrac{dE}{dk}$ when one of the related moduli k', $1/k$, $1/k'$, $\iota k/k'$, $\iota k'/k$ is substituted for k.

12. Deduce the expansions of K and E in powers of k by means of the equations
$$K=\int_0^{\frac{\pi}{2}}(1-k^2\sin^2\theta)^{-\frac{1}{2}}d\theta, \quad E=\int_0^{\frac{\pi}{2}}(1-k^2\sin^2\theta)^{\frac{1}{2}}d\theta.$$

13. From the equations of Ex. 12 find the values of $\dfrac{dK}{dk}$ and $\dfrac{dE}{dk}$.

CHAPTER X.

APPLICATIONS.

§ 87. The usefulness of the Elliptic Functions consists chiefly in this, that by means of them two surds of the form $(a + 2\beta x + \gamma x^2)^{\frac{1}{2}}$ can be rationalized at once. One such surd could be made rational by an algebraical substitution: thus $(1-x^2)^{\frac{1}{2}}$ becomes $(1-y^2)/(1+y^2)$ if $2y/(1+y^2)$ is put for x, and $(1+x^2)^{\frac{1}{2}}$ becomes $(1+y^2)/(1-y^2)$ if $2y/(1-y^2)$ is put for x; but generally speaking no rational algebraical or trigonometrical substitution will rationalize two such surds.

§ 88. Let the two surds be $s^{\frac{1}{2}}$ and $\sigma^{\frac{1}{2}}$ where
$$s = a + 2bx + cx^2, \qquad \sigma = a + 2\beta x + \gamma x^2,$$
We shall suppose the coefficients in s and σ to be real.

Also let $\qquad S = A + 2Bx + Cx^2$

where A, B, C are found from the equations
$$Ac - 2Bb + Ca = 0, \qquad A\gamma - 2B\beta + Ca = 0,$$
so that in fact $S = \begin{vmatrix} 1 & -x & x^2 \\ c & b & a \\ \gamma & \beta & a \end{vmatrix}.$

APPLICATIONS.

Let ξ, η be the two roots of the equation $S = 0$.

Then it is known that s and σ can both be expressed as sums of multiples of squares of $x - \xi, x - \eta$, and in fact it is easily verified, since

$$a + b(\xi + \eta) + c\xi\eta = 0,$$
$$a + \beta(\xi + \eta) + \gamma\xi\eta = 0,$$

that $\quad s(\xi - \eta) = (c\xi + b)(x - \eta)^2 - (c\eta + b)(x - \xi)^2,$

and $\quad \sigma(\xi - \eta) = (\gamma\xi + \beta)(x - \eta)^2 - (\gamma\eta + \beta)(x - \xi)^2.$

Also by tracing the rectangular hyperbolas

$$a + b(\xi + \eta) + c\xi\eta = 0,$$
$$a + \beta(\xi + \eta) + \gamma\xi\eta = 0,$$

each of which has the line $\xi = \eta$ for an axis, it is at once seen that the values of ξ and η which they furnish are real except when the line $\xi = \eta$ is the transverse axis in each, and each hyperbola has one vertex lying between those of the other. This is the case in which $s = 0$ and $\sigma = 0$ have both real roots, arranged so that one root of each falls between those of the other.

We see also that in the identity

$$s(\xi - \eta) = (c\xi + b)(x - \eta)^2 - (c\eta + b)(x - \xi)^2,$$

the product of the coefficients of the squares on the right is $-\{c^2\xi\eta + bc(\xi + \eta) + b^2\}$, that is $ac - b^2$.

Hence s is expressed as the sum of two squares if $s = 0$ has imaginary roots, as their difference if $s = 0$ has real roots. The same holds for σ.

If then ξ and η are real we may by the real rational substitution $y = (x - \eta)/(x - \xi)$, express $s^{\frac{1}{2}}$ and $\sigma^{\frac{1}{2}}$ in terms of y and two surds $(\pm 1 \pm \kappa^2 y^2)^{\frac{1}{2}}, (\pm 1 \pm \mu^2 y^2)^{\frac{1}{2}}.$

§ 89. Such a surd as $(-1 - \kappa^2 y^2)^{\frac{1}{2}}$ will be imaginary for all real values of y. The other cases we shall take in turn.

ELLIPTIC FUNCTIONS.

I. To rationalize $(1-\kappa^2 y^2)^{\frac{1}{2}}$, $(1-\mu^2 y^2)^{\frac{1}{2}}$. (Take $\kappa > \mu$.)

Put $\quad \kappa y = \operatorname{sn}\left(u, \dfrac{\mu}{\kappa}\right)$,

then $\quad (1-\kappa^2 y^2)^{\frac{1}{2}} = \operatorname{cn} u, \ (1-\mu^2 y^2)^{\frac{1}{2}} = \operatorname{dn} u.$

II. $(1-\kappa^2 y^2)^{\frac{1}{2}}$, $(1+\mu^2 y^2)^{\frac{1}{2}}$.

Put $\quad \kappa y = \operatorname{cn}\left\{u, \dfrac{\mu}{(\mu^2+\kappa^2)^{\frac{1}{2}}}\right\},$

then $(1-\kappa^2 y^2)^{\frac{1}{2}} = \operatorname{sn} u, \ (1+\mu^2 y^2)^{\frac{1}{2}} = \dfrac{1}{\kappa}(\mu^2+\kappa^2)^{\frac{1}{2}} \operatorname{dn} u.$

III. $(1+\kappa^2 y^2)^{\frac{1}{2}}$, $(1+\mu^2 y^2)^{\frac{1}{2}}$. (Take $\kappa > \mu$.)

Put $\quad \kappa y = \operatorname{sc}\left\{u, \dfrac{(\kappa^2-\mu^2)^{\frac{1}{2}}}{\kappa}\right\},$

then $\quad (1+\kappa^2 y^2)^{\frac{1}{2}} = \operatorname{nc} u, \ (1+\mu^2 y^2)^{\frac{1}{2}} = \operatorname{dc} u.$

IV. $(\kappa^2 y^2 - 1)^{\frac{1}{2}}$, $(1-\mu^2 y^2)^{\frac{1}{2}}$. Here κ must $> \mu$, or both surds cannot be real.

Put $\quad \mu y = \operatorname{dn}\left\{u, \dfrac{(\kappa^2-\mu^2)^{\frac{1}{2}}}{\kappa}\right\},$

then $\quad (1-\mu^2 y^2)^{\frac{1}{2}} = \dfrac{1}{\kappa}(\kappa^2-\mu^2)^{\frac{1}{2}} \operatorname{sn} u,$

$\quad (\kappa^2 y^2 - 1)^{\frac{1}{2}} = \dfrac{1}{\mu}(\kappa^2-\mu^2)^{\frac{1}{2}} \operatorname{cn} u.$

V. $(\kappa^2 y^2 - 1)^{\frac{1}{2}}$, $(1+\mu^2 y^2)^{\frac{1}{2}}$.

Put $\quad \kappa y = \operatorname{nc}\left\{u, \dfrac{\kappa}{(\kappa^2+\mu^2)^{\frac{1}{2}}}\right\},$

then $(\kappa^2 y^2 - 1)^{\frac{1}{2}} = \operatorname{sc} u, \ (1+\mu^2 y^2)^{\frac{1}{2}} = \dfrac{1}{\kappa}(\kappa^2+\mu^2)^{\frac{1}{2}} \operatorname{dc} u.$

VI. $(\kappa^2 y^2 - 1)^{\frac{1}{2}}$, $(\mu^2 y^2 - 1)^{\frac{1}{2}}$. (Take $\kappa > \mu$.)

Put $\quad\quad \mu y = \operatorname{ns}\left(u, \dfrac{\mu}{\kappa}\right),$

then $\quad (\kappa^2 y^2 - 1)^{\frac{1}{2}} = \dfrac{\kappa}{\mu}\operatorname{ds} u,\ (\mu^2 y^2 - 1)^{\frac{1}{2}} = \operatorname{cs} u.$

In each case the value of x is given in terms of u by substituting for y in

$$x = (y\xi - \eta)/(y - 1).$$

It hardly need be said that if ξ were infinite, we should put $y = x - \eta$, and then we could go on as before.

§ 90. If $\xi = \eta$, the process fails. But in that case s and σ have a common factor $x - \xi$.

Let $\quad s = (x - \xi)(cx + d), \quad \sigma = (x - \xi)(\gamma x + \delta).$

Put $\quad \dfrac{cx + d}{\gamma x + \delta} = y^2, \quad x = \dfrac{\delta y^2 - d}{c - \gamma y^2}$

Thus $\quad s(c - \gamma y^2)^2 = (\delta y^2 - d - \xi c + \xi \gamma y^2)(c\delta - d\gamma)y^2,$
$\quad\quad\ \ \sigma(c - \gamma y^2)^2 = (\delta y^2 - d - \xi c + \xi \gamma y^2)(c\delta - d\gamma),$

so that $s^{\frac{1}{2}}$ and $\sigma^{\frac{1}{2}}$ can be expressed by means of a single surd of the form $(A + By^2)^{\frac{1}{2}}$. This surd can again be rationalized by putting

$$y\sqrt{\pm \frac{B}{A}} = \frac{2m}{1 \mp m^2}.$$

Hence if $\xi = \eta$, the surds can be rationalized by an algebraical substitution.

§ 91. The above does not apply to the case when

$$s = c(x - d)(x - e), \quad \sigma = \gamma(x - \delta)(x - \epsilon),$$

d, δ, e, ϵ being real quantities in order of magnitude.

86   ELLIPTIC FUNCTIONS.

In this case put
$$\frac{x-d}{x-\delta}=y^2, \qquad x=\frac{\delta y^2-d}{y^2-1}.$$

Then $s = c(\delta-d)y^2\{(\delta-e)y^2-(d-e)\} \div (y^2-1)^2,$
$\sigma = \gamma(\delta-d)\ \{(\delta-\epsilon)y^2-(d-\epsilon)\} \div (y^2-1)^2.$

Thus $s^{\frac{1}{2}}$ and $\sigma^{\frac{1}{2}}$ are expressed by means of two surds only, and those of the form $(Ay^2+B)^{\frac{1}{2}}$, which we have already shown how to rationalize.

§ 92. It is easy to verify, and important to notice, that in each case $\dfrac{dx}{du}$ is a constant multiple of $s^{\frac{1}{2}}\sigma^{\frac{1}{2}}$.

§ 93. An expression of the form
$$ax^4+\beta x^3+\gamma x^2+\delta x+\epsilon\ (=X,\text{ say})$$
can always be expressed as the product of two real quadratic factors by the solution of a cubic equation. Hence any expression which is rational in x and $X^{\frac{1}{2}}$ can be rationalized by a substitution such as we have just discussed.

The exceptional case of § 91 need not arise. It will not be possible unless the roots of $X=0$ are all real. In that case there will be three ways of resolving X into real quadratic factors, and only one of the three will lead to the exceptional case.

If $a=0$, X becomes a cubic instead of a quartic; but by a linear substitution for x of the form
$$x=\frac{\kappa y+\lambda}{\mu y+\nu},$$
the expression is made rational in y and $Y^{\frac{1}{2}}$ where
$$Y=X(\mu y+\nu)^4,$$
so that Y is a quartic in y having $\mu y+\nu$ for one of

GEOMETRICAL APPLICATIONS. 87

its linear factors. Thus there is no real distinction between the cases of the cubic and the quartic.

§ 94. It must not be supposed that the rationalizing of these surds can only be accomplished by the particular substitutions which we have used. The number of substitutions that might be used is unlimited. We have tried to choose the simplest. The comparison of the different substitutions that would rationalize the same surd or pair of surds belongs to the theory of Transformations, which is beyond our limits.

APPLICATION IN THE INTEGRAL CALCULUS.

§ 95. When an expression has to be integrated which contains two surds, each the square root of a quadratic, or one surd which is the square root of a quartic, linear functions being counted as quadratic and cubic functions as quartic, then it follows from what we have proved that the integral can be expressed by means of the functions sn, cn, dn, E, Π.

For the subject of integration can be made a rational function of sn u, cn u, dn u by a properly chosen substitution, and such a function can be integrated as explained in Chapter IV.

GEOMETRICAL APPLICATIONS.

§ 96. The elliptic functions have an important use in the theory of curves, plane and twisted. This depends on the following theorem:—

The coordinates of any point on a curve whose deficiency is 1 can be expressed rationally by means of elliptic functions of a single parameter. (Compare Salmon, *Higher Plane Curves*, §§ 44, 366.)

Suppose the equation to the curve to be $U = 0$, and

that it has multiple points of orders, $k_1, k_2 \ldots$, its degree being m. Then the deficiency is

$$\tfrac{1}{2}(m-1)(m-2) - \Sigma \tfrac{1}{2}k(k-1),$$

and we have $\quad \Sigma \tfrac{1}{2}k(k-1) = \tfrac{1}{2}m(m-3)$.

Take a system of curves of the degree $m-2$, each having a point of order $k-1$, where $U=0$ has one of order k, and passing also through $m-2$ other fixed points on the curve.

The number of arbitrary coefficients in the equation to such a curve is $\tfrac{1}{2}(m+1)(m-2)$, and the number of conditions assigned is $\Sigma \tfrac{1}{2}k(k-1) + m - 2$, that is $\tfrac{1}{2}(m+1)(m-2) - 1$. Hence there will be one arbitrary coefficient left, and as all the equations to be satisfied by the coefficients were linear the equation to any curve of the system is $S + \lambda T = 0$, λ being the arbitrary coefficient and S, T determinate functions of the co-ordinates of the degree $m-2$, such that $S=0$, $T=0$ are two curves of the system.

Of the $m(m-2)$ intersections of the curves $U=0$, $S+\lambda T=0$, $\Sigma k(k-1) + m - 2$, that is $m^2 - 2m - 2$, are fixed. Thus only two depend on λ. Call these P and Q.

Let A be one of the $m-2$ fixed intersections of $S + \lambda T = 0$ with $U = 0$. Replace A by any other point A_1 taken at random on the curve. Then we have another system of curves $S_1 + \lambda_1 T_1 = 0$, whose intersections with $U=0$ are all fixed but two. Choose λ_1 so that P may be one of these and let Q_1 be the other.

Q_1 will not be the same as Q. For a curve of the degree $m-2$, satisfying all the conditions above prescribed for $S + \lambda T = 0$ except that of passing through A, and also passing through both P and Q, will be altogether fixed, and all its intersections with $U=0$ have been already specified but one. This one is A, and therefore it cannot be A_1. Hence Q_1 and Q are different.

The three equations $U=0$, $S+\lambda T=0$, $S_1+\lambda_1 T_1=0$ will therefore enable us to express the two coordinates of P *rationally* in terms of λ, λ_1, and also to eliminate those coordinates and find the relation between λ and λ_1.

When λ is given, there are two possible values for λ_1, found by substituting in $-S_1/T_1$ the coordinates of P and Q respectively. In the same way when λ_1 is given there are two possible values for λ. The equation connecting them must then be of the second degree in each, and may be written

$$\lambda_1^2(A\lambda^2+B\lambda+C)+\lambda_1(D\lambda^2+E\lambda+F)+G\lambda^2+H\lambda+I=0.$$

This equation may be solved for λ_1, the only irrational element being the square root of a quartic in λ. Hence this is the only irrational element in the expression of the coordinates of P in terms of λ, and it may be removed by a substitution for λ in terms of elliptic functions.

Thus the theorem is proved.

§ 97. If the curve is not plane, but twisted, we may suppose $S+\lambda T=0$, $S_1+\lambda_1 T_1=0$ to represent not curves but cones, of a degree lower by 2 than that of the curve. Take $U=0$ to be a cone with any vertex standing upon the curve and $S+\lambda T=0$ a cone with the same vertex, and having as a $(k-1)^{\text{ple}}$ edge any multiple edge of order k on $U=0$ and also having $m-2$ fixed edges in common with $U=0$.

$S_1+\lambda_1 T_1=0$ may then be a cone drawn in the same way with another vertex and we may ensure that Q_1 is not the same as Q as follows:—

Let the positions of P and Q when $\lambda=0$ be F and G. Through F and another point H draw a cone with the vertex that is proposed for $S_1+\lambda_1 T_1=0$ and satisfying those of the conditions that $S_1+\lambda_1 T_1=0$ must satisfy which are not at our disposal. Take the other $m-2$

simple intersections of this cone with the curve as defining the fixed edges of the system $S_1 + \lambda_1 T_1 = 0$. Then as G is not the same as H, Q_1 cannot in general be the same as Q.

The rest of the argument goes on as before, the two equations to the curve taking the place of the single equation $U = 0$.

The deficiency of a twisted curve is thus understood to mean that of its projection from an arbitrary point upon an arbitrary plane. In general the double points of the projection will not all be the projections of double points of the curve, but some at least will be the intersections with the plane of chords of the curve drawn from the vertex of projection.

§ 98. The simplest examples of curves of the kind in question are non-singular plane cubics, and among twisted curves the quartics which are the intersections of pairs of conicoids, and in particular sphero-conics.

If λ is the parameter of § 96, and u the elliptic argument, then it follows from § 92 that the coordinates are expressed rationally in terms of λ and $\dfrac{d\lambda}{du}$, which we may call λ', and λ'^2 is a rational quartic in λ. To each value of λ there correspond two values of u and two points on the curve the two corresponding values of λ' being equal with opposite signs.

§ 99. It may be proved that if a variable curve of any assigned degree meet the curve in points whose arguments are $u_1, u_2, \ldots, u_n$, then

$$u_1 + u_2 + \ldots + u_n = \text{a constant.}$$

For let $\phi_1 = 0$, $\phi_2 = 0$ be any two curves of the degree assigned. Then we can prove that for the intersections of the given curve with $\phi_1 + \mu\phi_2 = 0$, Σu is independent of μ.

In ϕ_1 and ϕ_2 substitute the values of the coordinates in terms of u, and let f_1, f_2 be the results of substitution.

Then u is given by the equation
$$f_1 + \mu f_2 = 0,$$
and
$$\frac{du}{d\mu} = -f_2 \div \left(\frac{df_1}{du} + \mu \frac{df_2}{du}\right).$$

Now f_1 and f_2 are rational functions of λ and λ', so that $f_2 \div (f_1 + \mu f_2)$ is also a rational function of them, say $\psi(\lambda, \lambda') \div \chi(\lambda, \lambda')$. Its denominator may be rationalized by writing it
$$\psi(\lambda, \lambda')\chi(\lambda, -\lambda') \div \chi(\lambda, \lambda')\chi(\lambda, -\lambda').$$

Thus since λ'^2 is rational in λ we may write
$$\frac{f_2}{f_1 + \mu f_2} = \frac{A + B\lambda'}{C},$$
A, B, C being rational functions of λ.

Let $\lambda_1, \lambda_2, \ldots, \lambda_n$ be the roots of the equation $C = 0$, corresponding to the values $u_1, u_2, \ldots, u_n$.

Then A/C and B/C may be resolved into partial fractions, there being an absolute term in the first case because A and C are of the same degree.

Hence we have an identity of the form
$$\frac{f_2}{f_1 + \mu f_2} = p_0 + \sum_{r=1}^{n} \frac{p_r + q_r \lambda'}{\lambda - \lambda_r}.$$

Now of the two points for which $\lambda = \lambda_r$, only one is generally to be taken, suppose that for which $\lambda' = \lambda_r'$. The left-hand side is therefore finite at the point for which $\lambda = \lambda_r$ and $\lambda' = -\lambda_r'$.

Making this substitution after multiplication by $\lambda - \lambda_r$, we find $p_r - q_r \lambda_r' = 0$.

Thus
$$\frac{f_2}{f_1 + \mu f_2} = p_0 + \sum_{r=1}^{n} \frac{q_r(\lambda' + \lambda_r')}{\lambda - \lambda_r}.$$

If, however, the point $(\lambda_r, -\lambda_r')$ is one of the intersections we must have $\lambda_s = \lambda_r$, $\lambda_s' = -\lambda_r'$ corresponding to u_s, another of the series $u_1, u_2, \ldots, u_n$. Then the equation $C = 0$ has only one root corresponding to the two arguments, and there is only one fraction $(p_r + q_r\lambda')/(\lambda - \lambda_r)$ for both.

But in this case the equation $p_r - q_r\lambda_r' = 0$ does not hold, and we write

$$\frac{p_r + q_r\lambda'}{\lambda - \lambda_r} = \frac{q_r\lambda_r' + p_r}{2\lambda_r'} \cdot \frac{\lambda' + \lambda_r'}{\lambda - \lambda_r} + \frac{q_r\lambda_s' + p_r}{2\lambda_s'} \cdot \frac{\lambda' + \lambda_s'}{\lambda - \lambda_s},$$

so that the final form is the same.

The identity

$$\frac{f_2}{f_1 + \mu f_2} = p_0 + \sum_{r=1}^{n} \frac{q_r(\lambda' + \lambda_r')}{\lambda - \lambda_r}$$

being thus proved to exist, we may find the value of q_r in the usual way, by multiplying by $\lambda - \lambda_r$ and putting $u = u_r$.

Thus $\quad q_r \cdot 2\lambda_r' = Lim_{u=u_r} \dfrac{f_2(\lambda - \lambda_r)}{f_1 + \mu f_2}$

$\quad\quad\quad\quad = $ value of $f_2\lambda' \div \left(\dfrac{df_1}{du} + \mu \dfrac{df_2}{du}\right)$,

when u_r is put for u,

$\quad\quad\quad\quad = -\lambda_r' du_r/d\mu$.

That is, $\quad q_r = -\tfrac{1}{2} du_r/d\mu$.

Now give u such a value that λ becomes infinite. Then λ' is infinite of a higher order; but as f_1 and f_2 are of the same degree, $f_2 \div (f_1 + \mu f_2)$ is finite. Thus

$$\Sigma q_r = 0,$$

and $\quad\quad\quad\quad \Sigma du_r/d\mu = 0$,

so that Σu_r is independent of μ.

AN EXAMPLE OF INTEGRATION.

Giving μ the two values 0 and ∞, we find that Σu_r is the same for the two curves $\phi_1 = 0$ and $\phi_2 = 0$. But these were taken to be any curves of the assigned degree. Hence the theorem is proved.

It will clearly hold also if the given curve is not plane and $\phi_1 = 0$, $\phi_2 = 0$ are any surfaces of the same degree.

§ 100. The facts proved in §§ 96-8 may be applied to integration. If y is a function of x, and the relation connecting them is the equation to a curve of deficiency 1, then any rational function of x and y may be expressed rationally by means of the functions sn, cn, dn of a single variable, and may be integrated with respect to x or y by means of these functions together with E and Π.

§ 101. Take, for instance, $\int (1-x^3)^{-\frac{2}{3}} dx$.

Put $$y = (1-x^3)^{\frac{1}{3}},$$
so that $$x^3 + y^3 = 1.$$

This is a cubic without singularity, so that the deficiency is 1.

Put $$x + y = z.^*$$
Then $$z^3 - 3xyz = 1,$$
$$xy = \frac{z^2}{3} - \frac{1}{3z}$$
$$(x-y)^2 = \frac{4}{3z} - \frac{z^2}{3}.$$

The radical is therefore $(4z - z^4)^{\frac{1}{2}}$.

The real quadratic factors of $z^4 - 4z$ are
$$z(z - 2^{\frac{2}{3}}) \text{ and } z^2 + 2^{\frac{2}{3}} z + 2^{\frac{4}{3}}.$$

---

* Here z takes the place of the λ of § 96, and the curves $S = 0$, $T = 0$ are respectively the straight line $x + y = 0$ and the line at infinity, the point of intersection of these two being clearly a point on the curve.

94   ELLIPTIC FUNCTIONS.

The roots of the equation
$$\begin{vmatrix} 2z-2^{\frac{2}{3}}, & -2^{\frac{2}{3}}z \\ 2z+2^{\frac{2}{3}}, & 2^{\frac{2}{3}}z+2^{\frac{1}{3}} \end{vmatrix}=0$$
are $2^{-\frac{1}{3}}(-1\pm\sqrt{3})$.
Hence we put
$$t=(2^{\frac{1}{3}}z-\sqrt{3}+1)/(2^{\frac{1}{3}}z+\sqrt{3}+1),$$
that is, $z=-\{t(\sqrt{3}+1)+(\sqrt{3}-1)\}\div 2^{\frac{1}{3}}(t-1).$
Then
$$(t-1)^2z(z-2^{\frac{2}{3}})=2^{\frac{1}{3}}\sqrt{3}(2+\sqrt{3})\{t^2-(2-\sqrt{3})^2\},$$
$$(t-1)^2(z^2+2^{\frac{2}{3}}z+2^{\frac{1}{3}})=2^{\frac{1}{3}}\cdot 3(t^2+1).$$

We therefore take
$$t=(2-\sqrt{3})\operatorname{cn}\left(u,\ \frac{\sqrt{3}-1}{2\sqrt{2}}\right),$$
using the substitution II of § 89, since the radical is $(4z-z^4)^{\frac{1}{2}}$, not $(z^4-4z)^{\frac{1}{2}}$.
Then
$$z=2^{-\frac{1}{3}}(\sqrt{3}-1)(1+\operatorname{cn} u)\div\{1-(2-\sqrt{3})\operatorname{cn} u\}$$
$$=2^{\frac{2}{3}}(1+\operatorname{cn} u)\div\{(\sqrt{3}+1)-(\sqrt{3}-1)\operatorname{cn} u\},$$
$$(t-1)^2z(z-2^{\frac{2}{3}})=-2^{\frac{1}{3}}\sqrt{3}(2-\sqrt{3})\operatorname{sn}^2 u,$$
$$(t-1)^2(z^2+2^{\frac{2}{3}}z+2^{\frac{1}{3}})=2^{\frac{1}{3}}\cdot 3(2-\sqrt{3})\operatorname{dn}^2 u.$$
Thus $(t-1)^4 z^2(x-y)^2=2^{\frac{5}{3}}3^{\frac{3}{2}}(2-\sqrt{3})^2\operatorname{sn}^2 u\,\operatorname{dn}^2 u,$
$$x-y=2^{\frac{1}{3}}3^{\frac{1}{4}}(2-\sqrt{3})\operatorname{sn} u\,\operatorname{dn} u$$
$$\div 2^{-\frac{1}{3}}(\sqrt{3}-1)(1+\operatorname{cn} u)\{1-(2-\sqrt{3})\operatorname{cn} u\}.$$
$$=2^{\frac{2}{3}}3^{\frac{1}{4}}\operatorname{sn} u\,\operatorname{dn} u$$
$$\div(1+\operatorname{cn} u)\{(\sqrt{3}+1)-(\sqrt{3}-1)\operatorname{cn} u\}.$$
Also $x+y=2^{\frac{2}{3}}(1+\operatorname{cn} u)\div\{(\sqrt{3}+1)-(\sqrt{3}-1)\operatorname{cn} u\}.$

AN EXAMPLE OF INTEGRATION.

From these equations x and y can be found at once.

Now if v be written for $\int (1-x^3)^{-\frac{2}{3}} dx$, we have, since
$$x^3 + y^3 = 1,$$
$$dv = \frac{dx}{y^2} = \frac{dy}{-x^2} = \frac{d(x+y)}{y^2 - x^2}.$$

But $\dfrac{d}{du}(x+y) = -2^{\frac{2}{3}} \operatorname{sn} u \operatorname{dn} u \{(\sqrt{3}+1) + (\sqrt{3}-1)\}$
$$\div \{(\sqrt{3}+1) - (\sqrt{3}-1)\operatorname{cn} u\}^2$$
$$= -3^{\frac{1}{2}} 2^{-\frac{2}{3}} (x-y)(x+y),$$

so that $\quad v = 2^{-\frac{2}{3}} \cdot 3^{\frac{1}{4}} \cdot u + \text{const.},$

that is to say,
$$\int \frac{dx}{(1-x^3)^{\frac{2}{3}}} = 2^{-\frac{2}{3}} 3^{\frac{1}{4}} \operatorname{cn}^{-1} \frac{(\sqrt{3}+1)\{x+(1-x^3)^{\frac{1}{3}}\} - 2^{\frac{2}{3}}}{(\sqrt{3}-1)\{x+(1-x^3)^{\frac{1}{3}}\} + 2^{\frac{2}{3}}} + \text{const.}$$
the modulus being $(\sqrt{3}-1)/2\sqrt{2}$.

§ 102. It should be noticed that when a is a constant, the equation connecting $\operatorname{sn} u$ and $\operatorname{sn}(u+a)$ is of the same doubly quadratic form as the one found between λ, λ_1 in § 96.

For the two values of $\operatorname{sn}(u+a)$ when $\operatorname{sn} u$ is given are $\operatorname{sn}(u+a)$ and $\operatorname{sn}(2K-u+a)$. Their sum is
$$2 \operatorname{sn} u \operatorname{cn} a \operatorname{dn} a \div (1 - k^2 \operatorname{sn}^2 u \operatorname{sn}^2 a),$$
and their product is
$$(\operatorname{sn}^2 u - \operatorname{sn}^2 a) \div (1 - k^2 \operatorname{sn}^2 u \operatorname{sn}^2 a).$$

Hence $\operatorname{sn}^2(u+a)\{1 - k^2 \operatorname{sn}^2 u \operatorname{sn}^2 a\}$
$$- 2\operatorname{sn}(u+a)\operatorname{sn} u \operatorname{cn} a \operatorname{dn} a + \operatorname{sn}^2 u - \operatorname{sn}^2 a = 0,$$

that is, $k^2 \operatorname{sn}^2 a \operatorname{sn}^2 u \operatorname{sn}^2(u+a) - \operatorname{sn}^2(u+a) - \operatorname{sn}^2 u$
$$+ 2 \operatorname{sn}(u+a) \operatorname{sn} u \operatorname{cn} a \operatorname{dn} a + \operatorname{sn}^2 a = 0.$$

The same holds for any other of the elliptic functions sn, cn, dn, sc, etc.

This suggests another way of integrating Euler's equation (§ 40) which was given by Cauchy.

Let $\phi(x, y) = 0$ be an equation of the second degree both in x and y, and let

$$\phi(x, y) = X_0 y^2 + 2X_1 y + X_2$$
$$= Y_0 x^2 + 2Y_1 x + Y_2.$$

Then
$$\frac{\partial \phi}{\partial x} = 2(Y_0 x + Y_1),$$

$$\frac{\partial \phi}{\partial y} = 2(X_0 y + X_1).$$

But since $\phi(x, y) = 0$ we have

$$(Y_0 x + Y_1)^2 = Y_1^2 - Y_0 Y_2 = Y, \text{ say,}$$
and $\quad (X_0 y + X_1)^2 = X_1^2 - X_0 X_2 = X, \text{ say.}$

Hence $\phi(x, y) = 0$ is an integral of the equation $X^{-\frac{1}{2}} dx + Y^{-\frac{1}{2}} dy = 0$, and X and Y are quartics in x and y respectively.

Also if in $\phi(x, y)$ the coefficients of $x^2 y$ and xy^2 are equal, as also those of x^2 and y^2, and those of x and y, then $\phi(x, y)$ will be symmetrical in x and y, and X will be the same function of x that Y is of y. Also the number of coefficients in ϕ is still one more than the number in X or Y so that if the coefficients of X and Y are known, $\phi = 0$ will contain one and only one arbitrary constant, and will be the complete primitive.

§ 103. If in a doubly quadratic equation connecting x and y we transform x or y or both by substitutions of the form $x = (e\xi + f)/(g\xi + h)$, the transformed equation is still of the same form in the new variables, though with different coefficients.

Now there are three arbitrary constants in such a transformation, and they may be so chosen as to make the transformed equation symmetrical, since symmetry is ensured if six coefficients are equal in pairs, namely those of x^2y, x^2, x to those of xy^2, y^2, y respectively.*

When the expression has been made symmetrical, x and y can be rationalized by a substitution for either in terms of elliptic functions, the two substitutions being of the same form and having the same modulus but different arguments. It follows however from the differential form of the equation that if u and v are the two arguments,

$$du = \pm dv, \quad u \pm v = \text{a constant.}$$

Hence transformations $x = \dfrac{e\xi + f}{g\xi + h}$, $y = \dfrac{a\eta + b}{c\eta + d}$ can be found such that ξ and η are the same function (sn, cn, dn, sc, etc.), with the same modulus, of arguments differing by a constant.

---

*With the notation of § 102, it may be proved that the anharmonic ratio of the roots of $X = 0$ is always the same as that of the roots of $Y = 0$.

For, by putting $xy = z$, $\phi(x, y)$ may be made a quadratic function of x, y and z, so that the two equations $xy - z = 0$, $\phi = 0$ represent a twisted quartic curve. The cone standing on this curve whose vertex is any point of it will be a cubic cone and the anharmonic ratio of the four tangent planes to it drawn through any one of its edges is a constant. (Salmon, *Higher Plane Curves*, § 167.) Thus if A, B, C, D are any four points on the curve the four tangent planes through AB have the same anharmonic ratio as those through BC, and these have the same as those through CD.

Now let AB, CD be the lines at infinity in the planes $x = 0$, $y = 0$ respectively, these being chords of the curve $xy = z$, $\phi = 0$. The equations $X = 0$, $Y = 0$ represent the two systems of tangent planes and the theorem follows. Another proof is given by Salmon (*Higher Plane Curves*, § 270).

It follows that by a linear transformation of x the roots of $X = 0$ can be made the same as those of $Y = 0$. This is the transformation wanted, for it may be verified that ϕ is symmetrical if the coefficients in X are proportional to those in Y. In carrying out this verification it is advisable to suppose X and Y reduced to their canonical form, in which the second and fourth terms are wanting. (See Salmon, *Higher Algebra*, § 203.)

98   ELLIPTIC FUNCTIONS.

§ 104. This applies to any case in which two parameters are connected by an algebraical relation, such that to each value of either there correspond two values of the other. There are two or three important cases of this which we shall now discuss.

In the first place, let P, Q be two points on a conic, such that the line joining them touches another fixed conic. If P is given there are two possible positions of Q, one on each of the tangents from P to the other conic. The relation between P and Q is reciprocal, and the coordinates of each may be expressed rationally in terms of a single parameter. Hence the parameters of the two points are connected by a doubly quadratic equation of the form we have been considering.

The same may be proved if the tangents at P and Q are to meet on another fixed conic, or if P and Q are to be conjugate points with respect to another fixed conic. It is in fact known that these three conditions are only the same stated in different ways.

§ 105. Jacobi has given a full discussion of the case when the two conics are circles, into which they can always be projected.

Take any four points A, a, β, B (Fig. 2), in order on a straight line, and on AB, $a\beta$ as diameters describe circles. Let the centres be Ω, O, the radii R, r, and let $O\Omega = \delta$.

Let P, Q be two points on the outer circle, such that PQ touches the inner circle at T. Let $P'T'Q'$ be a consecutive position of PTQ, meeting it in U.

Also write
$\theta = BAP, \quad \phi = BAQ, \quad \theta + d\theta = BAP', \quad \phi + d\phi = BAQ'$.
Then
$$B\Omega P = 2\theta,$$
$$P\Omega P' = 2d\theta,$$
$$PP' = 2Rd\theta,$$
$$QQ' = 2Rd\phi.$$

JACOBI'S CONSTRUCTION.

The angle $PUP' = QUQ'$,

and the angle $P'PQ = QQ'P'$.

Thus $\dfrac{PP'}{PU} = \dfrac{QQ'}{UQ'}$,

and in the limit $\dfrac{d\theta}{PT} = \dfrac{d\phi}{TQ}$.

But
$$PT^2 = OP^2 - OT^2$$
$$= R^2 + \delta^2 + 2R\delta \cos 2\theta - r^2,$$
$$TQ^2 = R^2 + \delta^2 + 2R\delta \cos 2\phi - r^2.$$

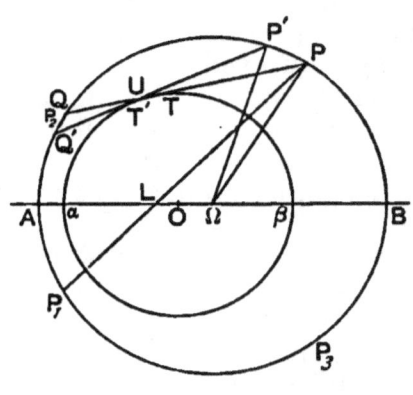

Fig. 2.

If then we write
$$k^2 = 4R\delta/\{(R+\delta)^2 - r^2\},$$
$$\sin \theta = \operatorname{sn}(u, k),$$
$$\sin \phi = \operatorname{sn}(v, k),$$

we have
$\cos \theta = \operatorname{cn} u,$
$\cos \phi = \operatorname{cn} v,$
$$PT = \{(R+\delta)^2 - r^2\}^{\frac{1}{2}} \operatorname{dn} u,$$
$$TQ = \{(R+\delta)^2 - r^2\}^{\frac{1}{2}} \operatorname{dn} v.$$

Also $\cos\theta\,d\theta = \operatorname{cn} u\,\operatorname{dn} u\,du,$
$d\theta = \operatorname{dn} u\,du.$
Thus $du = dv,$
$v - u = a,$ a constant.

§ 106. If now we put $\xi = \tan\theta$, $\eta = \tan\phi$, the co-ordinates of P and Q can be expressed rationally in terms of ξ and η respectively, and we can find the algebraical relation between ξ and η that follows from the equation $v - u = a$.

Take ΩB as axis of x, and a perpendicular to it from Ω as axis of y. Then the equation to PQ is

$$x\cos(\theta+\phi) + y\sin(\theta+\phi) = R\cos(\theta-\phi).$$

The perpendicular drawn to it from O is r. Hence

$$R\cos(\theta-\phi) + \delta\cos(\theta+\phi) = r,$$

that is, $R + \delta + (R-\delta)\xi\eta = r\sec\theta\sec\phi,$

$$(R+\delta)^2 + 2(R^2-\delta^2)\xi\eta + (R-\delta)^2\xi^2\eta^2 = r^2(1+\xi^2)(1+\eta^2).$$

Putting $r/(R+\delta) = \operatorname{cn}(a, k),$

the value of $\cos\phi$ when θ is 0, we find

$$(R-\delta)/(R+\delta) = \operatorname{dn}(a, k).$$

Thus $1 + 2\xi\eta\operatorname{dn} a + \xi^2\eta^2\operatorname{dn}^2 a = (1+\xi^2)(1+\eta^2)\operatorname{cn}^2 a.$

Solving the quadratic for η, we find

$$\eta = \frac{-\xi\operatorname{dn} a \pm \operatorname{sn} a\operatorname{cn} a(1+\xi^2)^{\frac{1}{2}}(1+k'^2\xi^2)^{\frac{1}{2}}}{\xi^2 k'^2\operatorname{sn}^2 a - \operatorname{cn}^2 a}.$$

As was to be expected, this is rationalized by the substitution $\xi = \operatorname{sc}(u, k)$, and becomes

$$\eta = \frac{\operatorname{sn} u\operatorname{cn} u\operatorname{dn} a \mp \operatorname{sn} a\operatorname{cn} a\operatorname{dn} u}{\operatorname{cn}^2 u\operatorname{cn}^2 a - k'^2\operatorname{sn}^2 a\operatorname{sn}^2 u},$$

so that $\operatorname{sc}(u+a) = \dfrac{\operatorname{sn} u\operatorname{cn} u\operatorname{dn} a + \operatorname{sn} a\operatorname{cn} a\operatorname{dn} u}{\operatorname{cn}^2 u\operatorname{cn}^2 a - k'^2\operatorname{sn}^2 a\operatorname{sn}^2 u},$

the lower sign being taken in order that the two sides may agree when $u = 0$. This is justifiable because a was found from its cn and dn, and therefore the sign of sn a is as yet undetermined.

The equation just found is one of the addition-formulae. Others may be written down at once from the figure. For instance,
$$PT + TQ = 2R\sin(\phi - \theta),$$
that is, $(R+\delta)\operatorname{sn} a \{\operatorname{dn} u + \operatorname{dn} v\}$
$$= (R+\delta)(1 + \operatorname{dn} a)(\operatorname{sn} v \operatorname{cn} u - \operatorname{sn} u \operatorname{cn} v),$$
or
$$\frac{\operatorname{sn}(u+a)\operatorname{cn} u - \operatorname{sn} u \operatorname{cn}(u+a)}{\operatorname{dn}(u+a) + \operatorname{dn} u} = \frac{\operatorname{sn} a}{1 + \operatorname{dn} a}.$$

§ 107. When the outer circle and AB, the axis of symmetry of the figure, are kept fixed, the quantities a and k depend on the position and size of the inner circle. It is of some importance to know under what circumstances the modulus k will be constant.

Now $\qquad k^2 = 4R\delta/\{(R+\delta)^2 - r^2\}.$

But if s is the distance from Ω of the radical axis of the two circles
$$s^2 - R^2 = (s-\delta)^2 - r^2,$$
and $\qquad 2s\delta = R^2 + \delta^2 - r^2,$
so that $\qquad s = 2R/k^2 - R.$

Hence if the inner circle vary so as always to have the same radical axis with the outer, the elliptic functions will have the same modulus. The quantity a is then the argument belonging to the other end of a chord of the outer circle drawn from B to touch the inner circle.

§ 108. An interesting case is that in which the inner circle has its radius zero, so that all the tangents to it pass through the inner limiting point of the coaxial system.

In that case cn $a = 0$, so that a is an odd multiple of K, if real. Let L be the limiting point. Then if PL produced meet the outer circle again in P_1, the argument $u + K$ belongs to the point P_1.

Thus $u + 2K$ belongs to P. It should, however, be noticed that when the argument u is increased by $2K$ in this way, θ is increased by π only, so that sn u and cn u have signs opposite to those they had before. The signs of BP and AP are in fact changed, because the positive direction of measurement has been changed in each case by a rotation through two right angles.

We have then
$$\operatorname{sn} u = BP/BA,$$
$$\operatorname{cn} u = AP/BA,$$
$$\operatorname{dn} u = LP/LB;$$

and, travelling along the arc PAP_1,
$$\operatorname{sn}(u+K) = BP_1/BA,$$
$$\operatorname{cn}(u+K) = -AP_1/BA,$$
$$\operatorname{dn}(u+K) = LP_1/LB.$$

Now $BP_1 = BA \sin BPL = BA \sin PBL \times BL/PL$
$$= PA \cdot BL/PL.$$

Thus $\operatorname{sn}(u+K) = \operatorname{cd} u.$

Also $AP_1 = PB \cdot AL/PL.$

Now $AL/BL = \operatorname{dn} K = k'.$

Thus $\operatorname{cn}(u+K) = -k' \operatorname{sd} u;$

and since $PL \cdot LP_1 = BL \cdot LA,$
$$\operatorname{dn}(u+K) = k' \operatorname{nd} u.$$

§ 109. The coaxial system of circles have a common self-polar triangle of which L is one angular point, the other two being L' the other limiting point and

the point at infinity in a direction perpendicular to
AB, which we may call M.

The figure shows that if $L'P$ and MP meet the
circle again in P_2 and P_3, the arguments belonging to
P_2 and P_3 are $K-u$ and $-u$ respectively, for P_2P_3
passes through L.

But since $\mathrm{sc}(2\iota K'-u) = \mathrm{sc}\, u$, every point on the
circle has two distinct (that is, not congruent) arguments belonging to it, and the second arguments
belonging to P_2, P_3 are respectively congruent to

$2\iota K' + K + u$ and $2\iota K' + u$ (mod. $2K$, $4\iota K'$).

It is now clear that if the inner circle in Jacobi's
construction is replaced by a circle of the same coaxial
sytem, but containing the other limiting point, then
the quantity a is not purely real but has its imaginary
part equal to an odd multiple of $2\iota K'$. If on the
other hand a is purely imaginary, its cn and dn are
real, so that the inner circle is to be replaced by a
real circle of the system, but one which contains the
original outer circle.

§ 110. By help of the foregoing we can answer the
following question: Can a polygon of an assigned
number of sides be inscribed in one given conic and
circumscribed to another?

Project the two conics into circles as before. Let u
be the argument of one angular point, $u + a$ that of
the next, then $u + 2a$ will be that of the third, and so
on, and if the polygon has n sides and is closed the
argument $u + na$ must belong to the first angular
point.

Hence $u + na \equiv u$ or $2\iota K' - u$ (mod. $2K$, $4\iota K'$).

Suppose first that

$$u + na \equiv 2\iota K' - u,$$

then
$$u + a \equiv 2\iota K' - u - (n-1)a,$$
$$u + 2a \equiv 2\iota K' - u - (n-2)a,\ \text{etc.},$$

so that the second angular point coincides with the nth, the third with the $(n-1)$th, and so on. Thus there is no proper polygon in this case.

If on the other hand we take $u + na \equiv u$ we find

$$a \equiv 0 \ (\text{mod. } 2K/n, 4\iota K'/n).$$

This condition does not assign any of the angular points, but only shows that unless the two conics are related in a particular way the problem has no solution. If the conics are so related, that is, if a has one of the values included in the formula $(2rK + 4s\iota K')/n$, then the value of u does not matter, and any point on the circumscribing conic may be taken as an angular point of the polygon.

ARCS OF CENTRAL CONICS.

§ 111. It is most likely known to the reader that the length of any elliptic arc can be expressed in terms of the coordinates of its ends by means of the elliptic functions sn, cn, dn, E, and that it is from this fact that the name "elliptic" arises.

The ellipse $x^2/a^2 + y^2/b^2 = 1$ is the locus of the point $(a\,\text{sn}\,u, b\,\text{cn}\,u)$ for different values of the argument u.

If S is the length of the arc measured from one end of the minor axis $(0, b)$ then S vanishes with u and

$$(dS/du)^2 = (a^2\text{cn}^2 u + b^2\text{sn}^2 u)\text{dn}^2 u$$
$$= a^2(1 - e^2\text{sn}^2 u)\text{dn}^2 u.$$

So far we have not assigned the value of k. If we take e for its value we have

$$dS/du = a\,\text{dn}^2 u$$

and
$$S = aE(u, e),$$

if
$$x = a\,\text{sn}(u, e),$$
$$y = b\,\text{cn}(u, e).$$

This expression holds equally well for the hyperbola, but it is not so useful, as the modulus of the elliptic functions is then greater than 1 and the point from which the arcs are measured is imaginary, b being imaginary.

§ 112. In the hyperbola $x^2/a^2 - y^2/b^2 = 1$ we may however put
$$y = b \operatorname{cs}(K-u) = bk' \operatorname{sc} u,$$
$$x = a \operatorname{ns}(K-u) = a \operatorname{dc} u.$$
so that u vanishes for the point $(a, 0)$.

If S is the length of the arc measured from this point we have
$$(dS/du)^2 = (a^2 k'^4 \operatorname{sc}^2 u + b^2 k'^2 \operatorname{dc}^2 u) \operatorname{nc}^2 u,$$
$$= b^2 k'^2 \operatorname{nc}^4 u,$$
if $a^2 k'^2 = b^2 k^2$, that is $k = 1/e$.

Thus $dS/du = bk' \operatorname{nc}^2 u$ if $k = 1/e$,
and $S = ae \{ \operatorname{sc} u \operatorname{dn} u + k'^2 u - Eu \}$.

§ 113. The equation
$$Eu + Ev - E(u+v) = k^2 \operatorname{sn} u \operatorname{sn} v \operatorname{sn}(u+v)$$
may be expected to furnish a geometrical theorem concerning arcs of a central conic.

We must first find what geometrical condition is expressed by such an equation as $u - v = t$, connecting the arguments u and v of two points on the ellipse. It will be more convenient to put
$$u = a + \beta, \quad v = a - \beta.$$
The tangents at u, v are then
$$\frac{x}{a} \operatorname{sn}(a \pm \beta) + \frac{y}{b} \operatorname{cn}(a \pm \beta) = 1,$$

and at their intersection we have

$$\frac{x}{a}\operatorname{sn}\alpha\operatorname{cn}\beta\operatorname{dn}\beta+\frac{y}{b}\operatorname{cn}\alpha\operatorname{cn}\beta = 1-k^2\operatorname{sn}^2\alpha\operatorname{sn}^2\beta,$$

$$\frac{x}{a}\operatorname{sn}\beta\operatorname{cn}\alpha\operatorname{dn}\alpha = \frac{y}{b}\operatorname{sn}\alpha\operatorname{sn}\beta\operatorname{dn}\alpha\operatorname{dn}\beta,$$

whence
$$x = a\operatorname{sn}\alpha\operatorname{dc}\beta,$$
$$y = b\operatorname{cn}\alpha\operatorname{nc}\beta.$$

Eliminating α, we have
$$x^2/a^2\operatorname{dc}^2\beta + y^2/b^2\operatorname{nc}^2\beta = 1.$$

Eliminating β, we have, since e is the modulus,
$$x^2/a^2e^2\operatorname{sn}^2\alpha - y^2/a^2e^2\operatorname{cn}^2\alpha = 1.$$

Each of these conics is confocal with the original one. Thus if $u \pm v$ is constant, the intersection of tangents at the points u, v traces a confocal conic.

§ 114. At a point on the tangent at u whose distance from the point of contact is z we have

$$\frac{x-a\operatorname{sn}u}{a\operatorname{cn}u} = \frac{y-b\operatorname{cn}u}{-b\operatorname{sn}u} = \frac{z}{a\operatorname{dn}u},$$

so that $x = a\operatorname{sn}u + z\operatorname{cd}u = a\operatorname{sn}u + z\operatorname{sn}(u+K)$,
$y = b\operatorname{cn}u + z\operatorname{cn}(u+K)$.

It is hence easily found that the lengths of the two tangents at $(\alpha \pm \beta)$ measured to their intersection are

$$a\operatorname{sc}\beta\operatorname{dn}\alpha\operatorname{dn}(\alpha\pm\beta).$$

Call these t_1, t_2. Then
$$t_1+t_2 = 2a\operatorname{sc}\beta\operatorname{dn}^2\alpha\operatorname{dn}\beta/(1-k^2\operatorname{sn}^2\alpha\operatorname{sn}^2\beta),$$
$$t_1-t_2 = -2ae^2\operatorname{sn}^2\beta\operatorname{sn}\alpha\operatorname{cn}\alpha\operatorname{dn}\alpha/(1-k^2\operatorname{sn}^2\alpha\operatorname{sn}^2\beta).$$

Now by the addition-formula for the function E
$$E(\alpha+\beta)-E\alpha-E\beta = -k^2\operatorname{sn}\alpha\operatorname{sn}\beta\operatorname{sn}(\alpha+\beta),$$
$$E(\alpha-\beta)-E\alpha+E\beta = k^2\operatorname{sn}\alpha\operatorname{sn}\beta\operatorname{sn}(\alpha-\beta),$$

GRAVES' THEOREMS.

and by addition and subtraction

$$E(a+\beta)+E(a-\beta)-2Ea$$
$$= -2k^2\operatorname{sn}^2\beta \operatorname{sn} a \operatorname{cn} a \operatorname{dn} a/(1-k^2\operatorname{sn}^2a \operatorname{sn}^2\beta)$$
$$= (t_1-t_2)/a,$$
$$E(a+\beta)-E(a-\beta)-2E\beta$$
$$= -2k^2\operatorname{sn}^2a \operatorname{sn} \beta \operatorname{cn} \beta \operatorname{dn} \beta/(1-k^2\operatorname{sn}^2a \operatorname{sn}^2\beta)$$
$$= (t_1+t_2)/a - 2\operatorname{sc} \beta \operatorname{dn} \beta.$$

If then $a+\beta$, $a-\beta$ are the arguments of the two points P and Q the tangents at which meet in T, and if B is the point from which the arcs are being measured, we have, when T traces a confocal ellipse, so that β is a real constant,

$$\text{arc } BP - \text{arc } BQ - TP - TQ = \text{a constant,}$$
or $$TP + TQ - \text{arc } PQ = \text{a constant;}$$

and when T traces a confocal hyperbola, so that a is a real constant,

arc $BP+$ arc $BQ - TP + TQ = $ a constant $=$ twice arc BR,

if R is the point of intersection of the hyperbola and ellipse between P and Q. Thus

$$TP - \text{arc } RP = TQ - \text{arc } RQ.$$

§ 115. This applies also to the hyperbola, but since in that case b is a pure imaginary the relation

$$TP + TQ - \text{arc } PQ = \text{a constant}$$

holds when T moves along a confocal hyperbola, and

$$TP - \text{arc } RP = TQ - \text{arc } RQ$$

when T moves along a confocal ellipse.

For geometrical proofs of these theorems, which are due to Dr. Graves, see Salmon's *Conic Sections* Chap. XIX.

It is noticeable that the system of confocal conics is the reciprocal of a system of coaxial circles with respect to one of the limiting points, so that this case is closely connected with that of §§ 107-110.

A CASE IN SPHERICAL GEOMETRY.

§ 116. Another case of a doubly quadratic relation between two parameters is afforded when an arc of a great circle moves on a sphere so as always to have its two ends on two fixed great circles, its length being constant.

Let PQ, $P'Q'$ be two consecutive positions of the movable arc, OPP', $OQ'Q$ the two fixed arcs (Fig. 3).

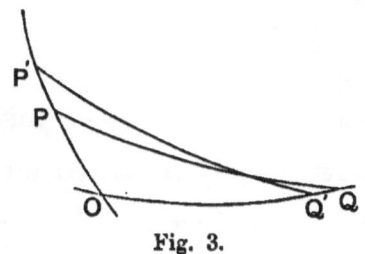

Fig. 3.

Let $OP = \theta$, $OP' = \theta + d\theta$, $OQ = \phi$,
$OQ' = \phi + d\phi$, $POQ = A$, $PQ = a$.

Then the integral equation connecting θ and ϕ is

$$\cos\theta \cos\phi + \cos A \sin\theta \sin\phi = \cos a.$$

To form the differential equation, since $PQ = P'Q'$, we have $PP' \cos OPQ = Q'Q \cos OQP$ in the limit, that is,

$$(1 - \sin^2 A \operatorname{cosec}^2 a \sin^2\phi)^{\frac{1}{2}} d\theta$$
$$+ (1 - \sin^2 A \operatorname{cosec}^2 a \sin^2\theta)^{\frac{1}{2}} d\phi = 0.$$

THE AMPLITUDE.

We may then put

$$\sin\theta = \operatorname{sn} u, \quad \cos\theta = \operatorname{cn} u, \quad \cos OQP = \operatorname{dn} u,$$
$$\sin\phi = \operatorname{sn} v, \quad \cos\phi = \operatorname{cn} v, \quad \cos OPQ = \operatorname{dn} v,$$

the modulus being $\sin A \operatorname{cosec} a$, and we have

$$du + dv = 0, \quad u + v = \text{constant} = w, \text{ say.}$$

Then w is the value of v given by supposing u and therefore θ to vanish, so that

$$\operatorname{sn} w = \sin a, \quad \operatorname{cn} w = \cos a, \quad \operatorname{dn} w = -\cos A,$$

and we have $\operatorname{cn} w = \operatorname{cn} u \operatorname{cn} v - \operatorname{dn} w \operatorname{sn} u \operatorname{sn} v$,
that is, $\operatorname{cn}(u+v) = \operatorname{cn} u \operatorname{cn} v - \operatorname{sn} u \operatorname{sn} v \operatorname{dn}(u+v)$.

This is one of the addition-formulae.

We have also

$$\cos\theta = \cos a \cos\phi + \sin a \sin\phi \cos OQP,$$

or $\operatorname{cn} u = \operatorname{cn}(u+v)\operatorname{cn} v + \operatorname{sn}(u+v)\operatorname{sn} v \operatorname{dn} u,$
and $\operatorname{cn} v = \operatorname{cn}(u+v)\operatorname{cn} u + \operatorname{sn}(u+v)\operatorname{sn} u \operatorname{dn} v.$

These three equations may be solved for

$$\operatorname{sn}(u+v), \quad \operatorname{cn}(u+v), \quad \operatorname{dn}(u+v).$$

If the modulus is to be real and less than unity and w real, we must have A obtuse and $a+A$ greater than two right angles. We may then write

$$\sin\theta = \operatorname{sn} u, \qquad \cos\theta = \operatorname{cn} u,$$
$$\sin\phi = \operatorname{sn}(w-u), \quad \cos\phi = \operatorname{cn}(w-u),$$

w being a constant.

§ 117. In this case we have

$$d\theta/du = \operatorname{dn} u \quad \text{or} \quad du/d\theta = (1 - k^2\sin^2\theta)^{-\frac{1}{2}}.$$

The function θ of u which satisfies this condition and vanishes with u was called by Jacobi the *ampli-*

tude of u, it being the upper limit on the right-hand side of the equation

$$u = \int_0^\theta (1-k^2\sin^2\theta)^{-\frac{1}{2}} d\theta.$$

It was also customary to write $\Delta\theta$ for $(1-k^2\sin^2\theta)^{\frac{1}{2}}$. Thus sn u, cn u, dn u were conceived as the sine, cosine and Δ of the amplitude of u, and in Jacobi's notation were written sin am u, cos am u, Δ am u, the amplitude θ being denoted by am u. The shorter notation, sn, cn, dn, was suggested by Gudermann.

The function am u is of no importance in the theory of elliptic functions, but it sometimes presents itself in the applications of the theory. In the case considered we may, for instance, write

$$\theta = \text{am } u, \quad \phi = \text{am}(w-u).$$

APPLICATIONS IN DYNAMICS. THE PENDULUM.

§ 118. There are certain problems in dynamics whose solution can be expressed by means of elliptic functions. The simplest is perhaps that of the motion of a pendulum.

The equation of motion is

$$l\ddot{\theta} = -g\sin\theta,$$

where θ is the inclination to the vertical of the plane through the axis of suspension and the centre of inertia and l is the length of the simple equivalent pendulum.

A first integral is found by multiplying by $\dot\theta$, it is

$$\tfrac{1}{2}l\dot\theta^2 = g(\kappa + \cos\theta) = g(1+\kappa - 2\sin^2\tfrac{1}{2}\theta),$$

κ being a constant. To integrate this put

$$\theta = 2\,\text{am}\{u,\, 2^{\frac{1}{2}}(1+\kappa)^{-\frac{1}{2}}\},$$

THE PENDULUM.

so that
$$\sin \tfrac{1}{2}\theta = \operatorname{sn} u,$$
$$\cos \tfrac{1}{2}\theta = \operatorname{cn} u,$$
$$1 + \kappa - 2\sin^2\tfrac{1}{2}\theta = (1+\kappa)\operatorname{dn}^2 u.$$
Then
$$\dot{u}^2 = (1+\kappa)g/2l,$$
and
$$u = t\{(1+\kappa)g/2l\}^{\tfrac{1}{2}} + \text{const.}$$

§ 119. Let A, B be the highest and lowest points of the circle described by the centre of inertia of the pendulum, P its position at any time, h its distance from the fixed horizontal axis, and let

$$(1+\kappa)g/2l = n^2.$$
Then
$$BP = 2h\operatorname{sn} nt,$$
$$AP = 2h\operatorname{cn} nt,$$

if the time is measured from the moment when P is at B.

If PY is the perpendicular drawn from P to a horizontal plane at a distance κh above the axis, that is, at the level of zero velocity, we have

$$PY = (1+\kappa)h\operatorname{dn}^2 nt.$$

Let BA, produced if necessary, meet this plane in C. Then let a circle be described having CY as its radical axis with the circle APB. The tangent from P to such a circle varies as $PY^{\tfrac{1}{2}}$, that is, as $\operatorname{dn} nt$. Hence the figure is the same as that in Jacobi's construction (§ 105 above).

§ 120. The application of the addition-formula will then give us the following theorem:—

The envelope of the line which joins the position of the centre of inertia at any time to its position at a fixed interval afterwards is a circle of the coaxial system which has for radical axis the line of zero velocity, and includes the circle described by the centre of inertia.

When the pendulum is performing complete revolutions $\kappa > 1$, and the elliptic functions have a modulus < 1. Thus if the fixed interval is half the whole time of revolution, the straight line joining the two positions will always pass through a fixed point, namely, the inner limiting point of the system of circles, whose depth below the radical axis is

$$h(\kappa^2 - 1)^{\frac{1}{2}}.$$

Further, the envelope of the line joining two variable positions of the centre of inertia, which are separated by equal intervals of time from any fixed position (one before, one after) is a circle of the same coaxial system; and if the revolutions are complete, and the fixed position is at a depth $h(\kappa^2 - 1)^{\frac{1}{2}}$ below the line of no velocity, the line always passes through the outer limiting point.

The velocity of the centre of inertia varies as the tangent drawn from it to any fixed circle of the coaxial system, or in the case of complete revolutions as the distance from either limiting point.

§ 121. In the case when the pendulum oscillates, $1 - \kappa$ is positive, so that the modulus of the elliptic functions is greater than unity. The expressions may be transformed by the usual formulae; putting $g = lm^2$, we have

$$BP = 2^{\frac{1}{2}}(1 + \kappa)^{\frac{1}{2}} h \operatorname{sn} mt,$$
$$AP = 2h \operatorname{dn} mt,$$

the modulus being now $2^{-\frac{1}{2}}(1 + \kappa)^{\frac{1}{2}}$. The velocity varies as $\operatorname{cn} mt$.

The general theorems derived above from the addition-formula still hold, the system of coaxial circles having now real intersections, namely, the extreme points reached in the oscillation. The limiting points are however imaginary, and the line joining positions separated by an interval of half the period is always horizontal, as is also that which joins two that are separated by equal intervals from the lowest. The coaxial circle, which is the envelope in this case, consists of the radical axis and the line at infinity, and the tangents to it pass through their intersection.

MOTION OF A RIGID BODY UNDER NO FORCES.

§ 122. Another interesting case is that of a rigid body in motion under the action of no forces. The centre of inertia will then move uniformly in a straight line or be at rest, and the motion of the body about its centre of inertia will be unaffected by the motion of the centre of inertia, which we will therefore suppose to be fixed.

Let $\omega_1, \omega_2, \omega_3$ be the angular velocities of the body at any time t about its three principal axes of inertia, and let A, B, C be the three corresponding moments of inertia, and suppose that they are in descending order of magnitude.

The equations of motion are then

$$A\dot{\omega}_1 = (B-C)\omega_2\omega_3,$$
$$B\dot{\omega}_2 = (C-A)\omega_3\omega_1,$$
$$C\dot{\omega}_3 = (A-B)\omega_1\omega_2.$$

The form of these suggests a substitution

$$\omega_1 = a\,\mathrm{cn}\,qt, \quad \omega_2 = -\beta\,\mathrm{sn}\,qt, \quad \omega_3 = \gamma\,\mathrm{dn}\,qt,$$

since the sign of $C-A$ is negative and opposite to those of $B-C$, $A-B$.

Making the substitution we have

$$Aqa = (B-C)\beta\gamma,$$
$$-Bq\beta = (C-A)\gamma a,$$
$$Cq\gamma k^2 = (A-B)a\beta.$$

Hence $\dfrac{Aa^2}{B-C} = \dfrac{B\beta^2}{A-C} = \dfrac{C\gamma^2 k^2}{A-B} = \dfrac{\alpha\beta\gamma}{q} = j^2$, say.

The equations are therefore satisfied if

$$\omega_1 = j\left(\frac{B-C}{A}\right)^{\frac{1}{2}} \operatorname{cn} q(t-t_0),$$

$$\omega_2 = -j\left(\frac{A-C}{B}\right)^{\frac{1}{2}} \operatorname{sn} q(t-t_0),$$

$$\omega_3 = \frac{j}{k}\left(\frac{A-B}{C}\right)^{\frac{1}{2}} \operatorname{dn} q(t-t_0),$$

where $\quad q = \dfrac{j}{k}\left\{\dfrac{(B-C)(A-C)(A-B)}{ABC}\right\}^{\frac{1}{2}},$

and the arbitrary constants of integration are j, the modulus k, and t_0.

The following two important equations are easily found either from the equations of motion or the integrals:—

$A\omega_1^2 + B\omega_2^2 + C\omega_3^2 = j^2(A - k'^2 B - k^2 C)/k^2 = T$, say,

$A^2\omega_1^2 + B^2\omega_2^2 + C^2\omega_3^2 = j^2(k^2 AB + k'^2 AC - BC)/k^2 = G^2$, say.

§ 123. Suppose now that (l, m, n) are the direction-cosines of a straight line fixed in space. We then find

$$\dot{l} = m\omega_3 - n\omega_2,$$
$$\dot{m} = n\omega_1 - l\omega_3,$$
$$\dot{n} = l\omega_2 - m\omega_1,$$

and ω_1, ω_2, ω_3, are now known functions of t. If these equations can be integrated the problem is completely solved.

The equations give

$$l\dot{l} + m\dot{m} + n\dot{n} = 0,$$

and therefore $\quad l^2 + m^2 + n^2 = \text{constant}.$

The value of this constant is known to be 1.

MOTION UNDER NO FORCES.   115

Also $A\omega_1 \dot{l} + B\omega_2 \dot{m} + C\omega_3 \dot{n}$
$$= l(C-B)\omega_2\omega_3 + m(A-C)\omega_3\omega_1 + n(B-A)\omega_1\omega_2$$
$$= -Al\dot{\omega}_1 - Bm\dot{\omega}_2 - Cn\dot{\omega}_3.$$

Hence $Al\omega_1 + Bm\omega_2 + Cn\omega_3 = K$, a constant.

This equation expresses that the line (l, m, n) makes a constant angle with that whose direction-cosines are $(A\omega_1/G, B\omega_2/G, C\omega_3/G)$ and shows therefore that this latter is fixed in space. It is easily found that the equations are actually satisfied if

$$l = A\omega_1/G, \quad m = B\omega_2/G, \quad n = C\omega_3/G.$$

§ 124. We may now simplify the problem by supposing the line (l, m, n) to be perpendicular to this known fixed line, that is by putting $K = 0$.

Let (λ, μ, ν) be the direction-cosines of another line perpendicular both to (l, m, n) and to

$$(A\omega_1/G, \quad B\omega_2/G, \quad C\omega_3/G),$$

so that $\quad G\lambda = Cm\omega_3 - Bn\omega_2$, etc.

Then since (λ, μ, ν) is also fixed in space we have

$$\dot{\lambda} = \mu\omega_3 - \nu\omega_2,$$

and $\quad l\dot{\lambda} - \lambda \dot{l} = (l\mu - \lambda m)\omega_3 - (l\nu - \lambda n)\omega_2$
$$= -(B\omega_2^2 + C\omega_3^2)/G$$
$$= -(T - A\omega_1^2)/G.$$

Also $\quad l^2 + \lambda^2 + A^2\omega_1^2/G^2 = 1.$

Hence $\dfrac{d}{dt} \arctan l/\lambda = G(T - A\omega_1^2)/(G^2 - A^2\omega_1^2).$

Thus $\quad l = \lambda \tan v,$

if $\quad v = \int \dfrac{G(T - A\omega_1^2)}{G^2 - A^2\omega_1^2} dt.$

This integral can be expressed in terms of the function Π, for the subject of integration is a known function of t.

116   ELLIPTIC FUNCTIONS.

Then l, m, n are given by the equations
$$Al\omega_1 + Bm\omega_2 + Cn\omega_3 = 0,$$
$$Gl\cot v - Cm\omega_3 + Bn\omega_2 = 0,$$
$$l^2 + m^2 + n^2 = 1,$$

or
$$\frac{l}{B^2\omega_2^2 + C^2\omega_3^2} = \frac{m}{CG\omega_3\cot v - AB\omega_1\omega_2}$$
$$= \frac{n}{-BG\omega_2\cot v - AC\omega_1\omega_3}$$
$$= \frac{1}{G\operatorname{cosec} v (B^2\omega_2^2 + C^2\omega_3^2)^{\frac{1}{2}}}.$$

To find λ, μ, ν we need only change v into $v + \frac{\pi}{2}$ in these expressions.

Referred to the three fixed axes, the direction-cosines of the principal axis of greatest moment are $(A\omega_1/G, l, \lambda)$, those of the mean axis $(B\omega_2/G, m, \mu)$, and those of the third principal axis $(C\omega_3/G, n, \nu)$. Hence the orientation of the body is completely determined at any time.

The actual value of v is found to be
$$v_0 + G(t-t_0)/C + \iota\Pi\{q(t-t_0), a\}$$
if $\operatorname{sn} a = \iota\{A(B-C)/C(A-B)\}^{\frac{1}{2}},$

the values of $\operatorname{cn} a$, $\operatorname{dn} a$ being both positive, as well as that of $-\iota \operatorname{sn} a$. v_0 is the value of v when $t = t_0$, and it varies according as different straight lines in the "Invariable Plane" are considered. a is a purely imaginary constant depending on the nature of the rigid body. k may be any real quantity. If it is numerically greater than unity the formulae may be reduced by the usual transformation to others in which the modulus is less than unity.

The values of $\arctan m/\mu$ and $\arctan n/\nu$ might have been found in terms of Π functions instead of that of arctan l/λ; the formulae thus found must however reduce to those we have by means of the formula for addition of parameters in the function Π.

A further discussion of the motion, with references, may be found in Routh's *Advanced Rigid Dynamics* (Chap. IV.).

ATTRACTION OF AN ELLIPSOID.

§ 125. The potential of a solid homogeneous ellipsoid at any point may also be conveniently expressed in terms of elliptic functions.

The expressions
$$x^2 = a^2 a'^2 a''^2/(a^2 - b^2)(a^2 - c^2),$$
$$y^2 = b^2 b'^2 b''^2/(b^2 - c^2)(b^2 - a^2),$$
$$z^2 = c^2 c'^2 c''^2/(c^2 - a^2)(c^2 - b^2),$$
for the coordinates of any point in terms of the semi-axes of the three conicoids of a confocal system that pass through it, suggest that we make x, y, z constant multiples of S, C, D respectively where
$$S = \operatorname{sn} u_1 \operatorname{sn} u_2 \operatorname{sn} u_3 = s_1 s_2 s_3, \text{ say,}$$
$$C = \operatorname{cn} u_1 \operatorname{cn} u_2 \operatorname{cn} u_3 = c_1 c_2 c_3,$$
$$D = \operatorname{dn} u_1 \operatorname{dn} u_2 \operatorname{dn} u_3 = d_1 d_2 d_3.$$
Since $k^2 k'^2 s_1^2 s_2^2 - k^2 c_1^2 c_2^2 + d_1^2 d_2^2 = k'^2$,
we have $k^2 k'^2 S^2/s_r^2 - k^2 C^2/c_r^2 + D^2/d_r^2 = k'^2$,
where $r = 1, 2$ or 3. This equation is the relation that connects S, C, D when u_r is a constant.

If then we put
$$x = l \cdot k^2 k' S, \quad y = l \cdot k^2 C, \quad z = l \cdot \iota D,$$
l being any constant, the locus of (x, y, z) when u_r is a constant will be a conicoid whose semi-axes are the square roots of
$$l^2 k^2 k'^2 s_r^2, \quad -l^2 k^2 k'^2 c_r^2, \quad -l^2 k'^2 d_r^2.$$

ELLIPTIC FUNCTIONS.

The differences of these quantities are constants, so that the different conicoids are all confocal.

§ 126. For an ellipsoid the imaginary part of u_r must be an odd multiple of $\iota K'$. It will be more convenient to have u_r real in this case; we therefore put $u_r + \iota K'$ for u_r throughout, and we have

$$x = lk'/kS, \quad y = l \cdot \iota D/kS, \quad z = -l \cdot C/S,$$

the squares of the semi-axes being now

$$l^2 k'^2 / s_r^2, \quad l^2 k'^2 d_r^2 / s_r^2, \quad l^2 k'^2 c_r^2 / s_r^2.$$

When u_r is constant and real, we now have an ellipsoid, when its real part is an odd multiple of K a hyperboloid of one sheet, and when its imaginary part is an odd multiple of $\iota K'$ a hyperboloid of two sheets. In other cases the surface $u_r = $ constant is imaginary.

Since then one surface of each kind passes through any point, we may suppose $u_1, \iota(u_2 - K), u_3 - \iota K'$ to be all real.

The semi-axes of the focal ellipse are found, by putting $u_r = K$, to be lk' and lk'^2; and, as l and k' are arbitrary, these may be made equal to any lengths whatever, so that any system of confocals whatever may be represented in this way.

§ 127. We must now transform the equation $\nabla^2 V = 0$, that is,

$$\frac{\partial^2 V}{\partial x^2} + \frac{\partial^2 V}{\partial y^2} + \frac{\partial^2 V}{\partial z^2} = 0.$$

Now, in the first place, if V is expressed in terms of S, C, D,

$$\frac{\partial V}{\partial u_1} = \frac{\partial V}{\partial S} c_1 d_1 s_2 s_3 - \frac{\partial V}{\partial C} s_1 d_1 c_2 c_3 - \frac{\partial V}{\partial D} k^2 s_1 c_1 d_2 d_3,$$

ATTRACTION OF AN ELLIPSOID.

$$\frac{\partial^2 V}{\partial u_1^2} = \frac{\partial^2 V}{\partial S^2}c_1^2 d_1^2 s_2^2 s_3^2 + \frac{\partial^2 V}{\partial C^2}s_1^2 d_1^2 c_2^2 c_3^2 + \frac{\partial^2 V}{\partial D^2}k^4 s_1^2 c_1^2 d_2^2 d_3^2$$

$$+ 2\frac{\partial^2 V}{\partial C \partial D}k^2 s_1^2 c_1 d_1 c_2 d_2 c_3 d_3 - 2\frac{\partial^2 V}{\partial D \partial S}k^2 c_1^2 s_1 d_1 s_2 d_2 s_3 d_3$$

$$- 2\frac{\partial^2 V}{\partial S \partial C}d_1^2 s_1 c_1 s_2 c_2 s_3 c_3 - \frac{\partial V}{\partial S}s_1 s_2 s_3 (d_1^2 + k^2 c_1^2)$$

$$- \frac{\partial V}{\partial C}c_1 c_2 c_3 (d_1^2 - k^2 s_1^2) - \frac{\partial V}{\partial D}k^2 d_1 d_2 d_3 (c_1^2 - s_1^2),$$

with symmetrical expressions for $\partial^2 V/\partial u_2^2$, $\partial^2 V/\partial u_3^2$.
Thus

$$(s_2^2 - s_3^2)\partial^2 V/\partial u_1^2 + (s_3^2 - s_1^2)\partial^2 V/\partial u_2^2 + (s_1^2 - s_2^2)\partial^2 V/\partial u_3^2$$
$$= -(s_2^2 - s_3^2)(s_3^2 - s_1^2)(s_1^2 - s_2^2)$$
$$\times [\partial^2 V/\partial S^2 + k'^2 \partial^2 V/\partial C^2 - k'^2 k^4 \partial^2 V/\partial D^2],$$

all the other terms disappearing.

If then we put $x = lk^2 k'S$, $y = lk^2 C$, $z = lD$, we have

$$-(s_2^2 - s_3^2)(s_3^2 - s_1^2)(s_1^2 - s_2^2)l^2 k^4 k'^2 \nabla^2 V$$
$$= (s_2^2 - s_3^2)\partial^2 V/\partial u_1^2$$
$$\quad + (s_3^2 - s_1^2)\partial^2 V/\partial u_2^2 + (s_1^2 - s_2^2)\partial^2 V/\partial u_3^2.$$

If now we change u_r into $u_r + \iota K'$, this becomes

$$-(s_2^2 - s_3^2)(s_3^2 - s_1^2)(s_1^2 - s_2^2)l^2 k'^2 \nabla^2 V \div s_1^2 s_2^2 s_3^2$$
$$= s_1^2(s_2^2 - s_3^2)\partial^2 V/\partial u_1^2$$
$$\quad + s_2^2(s_3^2 - s_1^2)\partial^2 V/\partial u_2^2 + s_3^2(s_1^2 - s_2^2)\partial^2 V/\partial u_3^2.$$

The equation $\nabla^2 V = 0$ is therefore to be replaced by

$$s_1^2(s_2^2 - s_3^2)\partial^2 V/\partial u_1^2$$
$$\quad + s_2^2(s_3^2 - s_1^2)\partial^2 V/\partial u_2^2 + s_3^2(s_1^2 - s_2^2)\partial^2 V/\partial u_3^2 = 0.$$

§ 128. Now it is known that the equipotential surfaces of a thin homogeneous homoeoid (shell bounded by two similar, similarly situated and concentric ellipsoids) are the confocal ellipsoids that lie outside it, that is, the surfaces represented by $u_1 = $ constant if our confocal system is that to which the surface of the shell belongs.

If V is the value of the potential it is a function of u_1 only, satisfying the equation just written, which now becomes
$$\partial^2 V/\partial u_1^2 = 0.$$
Hence $V = Qu_1 + R$, Q and R being constants.

Now V vanishes at infinity and at very distant points is in a ratio of equality to M/r where M is the mass of the shell and r the distance of the point from the centre.

Also at infinity $u_1 = 0$, and for small values of u_1 the surfaces may be regarded as spheres of radius lk'/s_1. Hence when u_1 is small we have
$$Ms_1/lk' = Qu_1 + R,$$
that is, $\qquad R = 0, \qquad Q = M/lk'.$

The potential of the homoeoidal shell is therefore
$$Mu_1/lk'.$$

§ 129. If now we have a homogeneous solid ellipsoid whose semi-axes in descending order of magnitude are a, b, c and whose density is ρ, it may be divided up into thin homoeoidal shells, to each of which the foregoing will apply. To get the different shells we need only suppose l to vary in the above expression from 0 to $\frac{a}{k'} \operatorname{sn} v_1$, its value for the outside surface, v_1 being the constant value of u_1 for the outside surface referred to its own system of confocals.

The sum of the volumes of all the shells up to any value of l is
$$\tfrac{4}{3}\pi l^3 k'^3 \operatorname{cn} v_1 \operatorname{dn} v_1/\operatorname{sn}^3 v_1,$$
so that we substitute for M the expression
$$4\pi\rho l^2 dl \cdot k'^3 \operatorname{cn} v_1 \operatorname{dn} v_1/\operatorname{sn}^3 v_1,$$

ATTRACTION OF AN ELLIPSOID.

which is the differential of this with respect to l multiplied by ρ. The potential of the solid ellipsoid at an external point is therefore

$$\frac{4\pi\rho k'^2 \operatorname{cn} v_1 \operatorname{dn} v_1}{\operatorname{sn}^3 v_1} \int_0^{\frac{a}{k'}\operatorname{sn} v_1} u_1 l \, dl,$$

and u_1 is given as a function of l by the equation

$$x^2 \operatorname{sn}^2 u_1 + y^2 \operatorname{sd}^2 u_1 + z^2 \operatorname{sc}^2 u_1 = l^2 k'^2,$$

(x, y, z) being the coordinates of the external point. We find at once

$$k'^2 l \, dl = (x^2 s_1 c_1 d_1 + y^2 s_1 c_1 / d_1^3 + z^2 s_1 d_1 / c_1^3) du_1.$$

Thus if now we write u_1 for the value of u_1 at (x, y, z) in the system of confocals to which the outside surface belongs we have for the potential

$$\frac{4\pi\rho \operatorname{cn} v_1 \operatorname{dn} v_1}{\operatorname{sn}^3 v_1} \int_0^{u_1} u\{x^2 + y^2 \operatorname{nd}^4 u + z^2 \operatorname{nc}^4 u\} \operatorname{sn} u \operatorname{cn} u \operatorname{dn} u \, du$$

$$= \frac{2\pi\rho \operatorname{cn} v_1 \operatorname{dn} v_1}{\operatorname{sn}^3 v_1} \Bigg[u_1(x^2 \operatorname{sn}^2 u_1 + y^2 \operatorname{sd}^2 u_1 + z^2 \operatorname{sc}^2 u_1)$$

$$- \int_0^{u_1} (x^2 \operatorname{sn}^2 u + y^2 \operatorname{sd}^2 u + z^2 \operatorname{sc}^2 u) du \Bigg].$$

Also by definition of u_1,

$$x^2 \operatorname{sn}^2 u_1 + y^2 \operatorname{sd}^2 u_1 + z^2 \operatorname{sc}^2 u_1 = a^2 \operatorname{sn}^2 v_1,$$

and

$$\int_0^{u_1} \operatorname{sn}^2 u \, du = \frac{u_1}{k^2} - \frac{1}{k^2} E u_1,$$

$$\int_0^{u_1} \operatorname{sd}^2 u \, du = -\frac{1}{k'^2} \frac{\operatorname{sn} u_1 \operatorname{cn} u_1}{\operatorname{dn} u_1} - \frac{u_1}{k^2} + \frac{1}{k^2 k'^2} E u_1,$$

$$\int_0^{u_1} \operatorname{sc}^2 u \, du = \frac{1}{k'^2} \frac{\operatorname{sn} u_1 \operatorname{dn} u_1}{\operatorname{cn} u_1} - \frac{1}{k'^2} E u_1.$$

122   ELLIPTIC FUNCTIONS.

Hence the potential
$$= \frac{2\pi\rho \operatorname{cn} v_1 \operatorname{dn} v_1}{\operatorname{sn}^3 v_1}\bigg[u_1\{a^2\operatorname{sn}^2 v_1 - (x^2-y^2)/k^2\}$$
$$+ \frac{\operatorname{sn} u_1}{k'^2\operatorname{cn} u_1 \operatorname{dn} u_1}(y^2\operatorname{cn}^2 u_1 - z^2\operatorname{dn}^2 u_1)$$
$$+ \frac{Eu_1}{k^2 k'^2}(k'^2 x^2 - y^2 + k^2 z^2)\bigg].$$

Here
$$k^2 = (a^2-b^2)/(a^2-c^2),$$
$$k'^2 = (b^2-c^2)/(a^2-c^2),$$
$$\operatorname{dn} v_1 = b/a,$$
$$\operatorname{cn} v_1 = c/a,$$
$$\operatorname{sn} v_1 = (a^2-c^2)^{\frac{1}{2}}/a,$$

and u_1 is the least real positive argument that satisfies the equation
$$x^2\operatorname{sn}^2 u_1 + y^2\operatorname{sd}^2 u_1 + z^2\operatorname{sc}^2 u_1 = a^2 - c^2.$$

If the point (x, y, z) lies on the outer surface, we have $u_1 = v_1$.

§ 130. If the point (x, y, z) lies inside the ellipsoid, the above formula ceases to hold. We may however describe through (x, y, z) a similar, similarly situated and concentric surface, and use the above expression for the volume contained.

If λa, λb, λc are the semi-axes of this one, its potential is
$$\frac{2\pi\rho \operatorname{cn} v_1 \operatorname{dn} v_1}{\operatorname{sn}^3 v_1}\bigg[v_1\{\lambda^2 a^2 \operatorname{sn}^2 v_1 - (x^2-y^2)/k^2\}$$
$$+ \frac{\operatorname{sn} v_1}{k'^2 \operatorname{cn} v_1 \operatorname{dn} v_1}(y^2\operatorname{cn}^2 v_1 - z^2\operatorname{dn}^2 v_1)$$
$$+ \frac{Ev_1}{k^2 k'^2}(k'^2 x^2 - y^2 + k^2 z^2)\bigg].$$

EXAMPLES X.   123

We have then to deal with the outer shell. This may be divided into thin homoeoids as before. The potential of each is the same at all points inside it, and equal to

$$4\pi\rho l\, dl \cdot k'^2 v_1 \operatorname{cn} v_1 \operatorname{dn} v_1 / \operatorname{sn}^3 v_1.$$

This is to be integrated with respect to l between the limits $\lambda a \operatorname{sn} v_1/k'$ and $a \operatorname{sn} v_1/k'$, and added to the potential of the inner part.

The integral is

$$2\pi\rho a^2 (1-\lambda^2) \cdot v_1 \operatorname{cn} v_1 \operatorname{dn} v_1 / \operatorname{sn} v_1,$$

and the potential of the whole ellipsoid at an internal point (x, y, z) is found to be

$$2\pi\rho \frac{\operatorname{cn} v_1 \operatorname{dn} v_1}{\operatorname{sn}^3 v_1} \bigg[v_1 \{ a^2 \operatorname{sn}^2 v_1 - (x^2 - y^2)/k^2 \}$$

$$+ \frac{\operatorname{sn} v_1}{k'^2 \operatorname{cn} v_1 \operatorname{dn} v_1} (y^2 \operatorname{cn}^2 v_1 - z^2 \operatorname{dn}^2 v_1)$$

$$+ \frac{E v_1}{k^2 k'^2} (k'^2 x^2 - y^2 + k^2 z^2) \bigg].$$

The expression is the same as for an external point, but that the constant v_1 takes the place of the variable u_1.

EXAMPLES ON CHAPTER X.

1. Prove that $(1 - 2x^2 \cos 2a + x^4)^{\frac{1}{2}}$ can be rationalized by putting

$$x + \frac{1}{x} = 2 \operatorname{ns}(2u, \cos a),$$

and that then

$$x - \frac{1}{x} = -2 \operatorname{cs}(2u, \cos a),$$

$$\left(x^2 - 2\cos 2a + \frac{1}{x^2}\right)^{\frac{1}{2}} = 2\,\mathrm{ds}(2u,\cos a),$$

$$u = \int_0^x (1 - 2x^2\cos 2a + x^4)^{-\frac{1}{2}} dx.$$

2. Discuss the spherical figure of §116 in the case when $\sin A > \sin a$ and show that in that case we may put

$$\sin OPQ = \mathrm{sn}(u, \sin a \operatorname{cosec} A),$$
$$\sin OQP = \mathrm{sn}(w - u, \sin a \operatorname{cosec} A),$$

where $\pi - A = \mathrm{am}\, w$.

3. If $\cos\theta = \cos\beta\,\mathrm{dn}\,u$, $\tan\phi = \dfrac{\sin a}{\sin\beta}\,\mathrm{sc}\,u$,

where $\cos a = k'\cos\beta$, prove that the point whose polar coordinates are (R, θ, ϕ), R being a constant, traces a sphero-conic whose semi-axes are a, β and that the area of a central sector of this sphero-conic is

$$R^2 \sin a \sin\beta \int \frac{\mathrm{dn}\,u\,du}{1 + \cos\beta\,\mathrm{dn}\,u}.$$

4. Prove that the chord joining the points $u \pm a$ on this sphero-conic touches the sphero-conic whose equation is

$$\cot^2\theta\,\mathrm{cn}^2 a = \cot^2\beta\,\mathrm{dn}^2 a\,\cos^2\phi + \cot^2 a\,\sin^2\phi,$$

and that this has the same cyclic arcs as the former one.

5. Show that the sector bounded by the semi-diameters to the points $u \pm a$ differs from

$$2R^2 \arctan \frac{\mathrm{dn}\,u(\mathrm{sn}^2 a + \mathrm{cn}^2 a \cos^2\beta) + \mathrm{dn}\,a \cos\beta}{\mathrm{sn}\,a\,\mathrm{cn}\,a \sin a \sin\beta}$$

by a quantity independent of u.

Prove also that the area of the spherical triangle

formed by these two semi-diameters and the chord joining the points $u \pm a$ is
$$2R^2 \arctan \frac{\sin a \sin \beta \operatorname{sn} a \operatorname{cn} a \operatorname{dn} u}{1 - \operatorname{sn}^2 a \sin^2 a + \operatorname{dn} u \operatorname{dn} a \cos \beta},$$
and that the area of the segment cut off by this chord is independent of u.

6. In the same sphero-conic
$$(\cot^2\theta = \cot^2 a \sin^2\phi + \cot^2\beta \cos^2\phi)$$
prove that by the substitution
$$\tan \phi = \tan a \cot \beta \sin a \operatorname{cosec} \beta \operatorname{cs}(u, k),$$
where $\quad k' = \sin \beta \operatorname{cosec} a,$
the expression for the arc is reduced to
$$R \tan a \tan \beta \sin \beta \int \frac{du}{\tan^2\beta \operatorname{sn}^2 u + \tan^2 a \operatorname{cn}^2 u}.$$

7. Prove that at the intersection of tangents to this sphero-conic at the points $u \pm a$ (as in Ex. 6, not Ex. 3)
$$\frac{\cot \theta}{\operatorname{cn} a \operatorname{dn} u} = \frac{\cot a \sin \phi}{\operatorname{dn} a \operatorname{cn} u} = \frac{\cot \beta \cos \phi}{k' \operatorname{sn} u},$$
and that as u varies this point traces the confocal sphero-conic
$$\cot^2\theta \operatorname{nc}^2 u = \cot^2 a \sin^2\phi \operatorname{nd}^2 u + \cot^2\beta \cos^2\phi.$$

8. The length of the tangent at $u + a$ in the last example is
$$R \arctan \frac{\tan a \tan \beta \operatorname{sn} a \sin \beta}{\tan^2 a \operatorname{cn} u \operatorname{cn}(u+a) \operatorname{dn} a + \tan^2\beta \operatorname{sn} u \operatorname{sn}(u+a)}.$$
Find the differential coefficient of this expression with respect to u in the form
$$R \tan a \tan \beta \sin \beta \left[\frac{1}{\tan^2 a \operatorname{cn}^2(u+a) + \tan^2\beta \operatorname{sn}^2(u+a)} - \frac{1}{\tan^2 a \operatorname{cn}^2 u + \tan^2\beta \operatorname{sn}^2 u} \right],$$

and prove that the sum of the two tangents exceeds the intercepted arc by a quantity independent of u.

(Compare Salmon, *Geometry of Three Dimensions*, § 252.)

9. Verify that when
$$\operatorname{sn}^2 u \tan^2 \beta + \operatorname{cn}^2 u \tan^2 a = 0,$$
then
$$\operatorname{sn} u = \pm \frac{1}{k} \cos \beta, \quad \operatorname{cn} u = \pm \frac{\iota}{k} \sin \beta \cot a, \quad \operatorname{dn} u = \pm \sin \beta,$$
and the above expression for the length of the tangent becomes $R \arctan \pm \iota$.

10. Prove that the following equations give the motion of a heavy particle constrained to move on a fixed smooth spherical surface:—
$$\cos \theta = \cos a \operatorname{sn}^2 \omega t + \cos \beta \operatorname{cn}^2 \omega t,$$
$$\phi = \tfrac{1}{4} n \left\{ \int_0^{\omega t} \frac{du}{\sin^2 \tfrac{1}{2} a \operatorname{sn}^2 u + \sin^2 \tfrac{1}{2} \beta \operatorname{cn}^2 u} + \int_0^{\omega t} \frac{du}{\cos^2 \tfrac{1}{2} a \operatorname{sn}^2 u + \cos^2 \tfrac{1}{2} \beta \operatorname{cn}^2 u} \right\},$$

where θ is the angular distance of the particle from the lowest point, a, β are the greatest and least values taken by θ during the motion, ϕ is the angle made by the vertical plane through the centre and the particle at time t with its initial position, t being measured from a time when $\theta = \beta$, l is the radius of the sphere, and
$$k^2 = (\cos^2 \beta - \cos^2 a)/(1 + \cos^2 \beta + 2 \cos a \cos \beta),$$
$$\omega = \frac{1}{k} \left\{ \frac{g(\cos \beta - \cos a)}{2l} \right\}^{\frac{1}{2}},$$
$$n^2 = 4 \sin^2 a \sin^2 \beta / (1 + \cos^2 \beta + 2 \cos a \cos \beta).$$

EXAMPLES X.

11. Reduce the above value of ϕ to the form
$$n\{\omega t \operatorname{cosec}^2\beta + \tfrac{1}{4}\operatorname{sc} a \operatorname{cosec} \tfrac{1}{2}a \operatorname{cosec} \tfrac{1}{2}\beta \Pi(\omega t, a)$$
$$+ \tfrac{1}{4}\operatorname{sc} b \operatorname{sec} \tfrac{1}{2}a \operatorname{sec} \tfrac{1}{2}\beta \Pi(\omega t, b)\},$$
where
$$\operatorname{dn} a = \sin \tfrac{1}{2} a / \sin \tfrac{1}{2}\beta,$$
$$\operatorname{dn} b = \cos \tfrac{1}{2} a / \cos \tfrac{1}{2}\beta.$$
What is the general character of the motion?

12. On a curve of deficiency 1 and degree n, the sum of the arguments of its intersections with a curve of degree m is σ. Show that if $n > 3$ the fact of the sum of the arguments of mn points on the curve being σ does not ensure that the points lie on an m^{ic}, but that if $n = 3$ this condition is enough.

13. If the curve of intersection of two conicoids is projected from any point of itself on any plane, the projections will all be projections of the same plane cubic.

[The anharmonic ratio of the four tangents drawn to any of the cubics from a point on itself is the same for all. It may be expressed as a function of the elliptic modulus.]

14. Verify that the expressions found (§§ 129, 130) for the potential of an ellipsoid satisfy Laplace's and Poisson's equations, and find the components of the attraction at any point.

15. In Jacobi's coaxial circle figure (Fig. 2, § 105), prove that when $u = \iota K'$, O is at B, and when $u = K + \iota K'$, at A. In general when O lies between L and L', so that the variable circle is imaginary, the real part of u is an odd multiple of K.

16. The arguments of the circular points at infinity are $\pm \iota K'$, and of the other common points of the coaxial system $K \pm \iota K'$.

17. If l, m, n are in descending order of magnitude show that the two ends of a chord of the circle $x^2 + y^2 = m^2$ which touches the ellipse $x^2/l^2 + y^2/n^2 = 1$

have for their coordinates $km\,\mathrm{sn}(u\pm a)$, $m\,\mathrm{dn}(u\pm a)$, where

$$k^2 = \frac{l^2(m^2-n^2)}{m^2(l^2-n^2)}, \quad \mathrm{cn}\,a = \frac{n}{l}, \quad \mathrm{dn}\,a = \frac{n}{m},$$

and u is a variable parameter.

18. If $x+\iota y = \mathrm{sn}(u+\iota v)$, the points on the curves $u=\mathrm{const.}$, $v=\mathrm{const.}$ at which the tangents are parallel to the axes of coordinates, lie either on one of those axes or on a rectangular hyperbola whose axes they are. (See Appendix A.)

19. If

$$x+\iota y = \mathrm{sn}^2(u+\iota v)$$

or $\mathrm{cn}^2(u+\iota v)$ or $\mathrm{dn}^2(u+\iota v)$ or $\wp(u+\iota v)$,

the curves $u=\mathrm{const.}$, $v=\mathrm{const.}$ are confocal Cartesian ovals, and for one value of each the oval becomes a circle. Distinguish between the outer and inner ovals. (Greenhill.)

20. Examine the curves $u=\mathrm{const.}$, $v=\mathrm{const.}$ when

$$x+\iota y = \mathrm{sn}(u+\iota v)\,\mathrm{dc}(u+\iota v).$$

[The distances of the point (x, y) from the points $(\pm k, \pm k')$ are found to satisfy two linear relations. Hence the curves are bicircular quartics having these points for foci. In the particular cases when $u = \pm \tfrac{1}{2}K$, or $v = \pm \tfrac{1}{2}K'$ they become arcs of the circle $x^2+y^2=1$.]

APPENDIX A.

THE GRAPHICAL REPRESENTATION OF ELLIPTIC FUNCTIONS.

§ 131. The nature of the elliptic functions unfits them for representation by a linear graph as in the case of functions of a real variable. We may however get some idea of their variations by means of Argand's Diagram.

Let
$$x + \iota y = \operatorname{sn}(u + \iota v),$$

x, y, u, v being real, and let us examine the curves $u = $ constant, $v = $ constant; we need not consider values of u outside the limits $\pm 2K$ or of v outside $\pm K'$.

Call the point (x, y) P and the points $(1, 0)$, $(-1, 0)$, $(1/k, 0)$, $(-1/k, 0)$, A, B, C, D respectively. Then

$$AP^2 = \{1 - \operatorname{sn}(u + \iota v)\}\{1 - \operatorname{sn}(u - \iota v)\}$$
$$= (\operatorname{cn} \iota v - \operatorname{dn} \iota v \operatorname{sn} u)^2 / (1 - k^2 \operatorname{sn}^2 u \operatorname{sn}^2 \iota v),$$
$$BP^2 = (\operatorname{cn} \iota v + \operatorname{dn} \iota v \operatorname{sn} u)^2 / (1 - k^2 \operatorname{sn}^2 u \operatorname{sn}^2 \iota v),$$
$$k^2 CP^2 = \{1 - k \operatorname{sn}(u + \iota v)\}\{1 - k \operatorname{sn}(u - \iota v)\}$$
$$= (\operatorname{dn} \iota v - k \operatorname{cn} \iota v \operatorname{sn} u)^2 / (1 - k^2 \operatorname{sn}^2 u \operatorname{sn}^2 \iota v),$$
$$k^2 DP^2 = (\operatorname{dn} \iota v + k \operatorname{cn} \iota v \operatorname{sn} u)^2 / (1 - k^2 \operatorname{sn}^2 u \operatorname{sn}^2 \iota v).$$

Hence $\dfrac{BP - AP}{\operatorname{dn} \iota v \operatorname{sn} u} = \dfrac{BP + AP}{\operatorname{cn} \iota v} = \dfrac{DP - CP}{\operatorname{cn} \iota v \operatorname{sn} u} = \dfrac{k(DP + CP)}{\operatorname{dn} \iota v}.$

Thus the locus when v is a constant is given by
$$BP - AP = (DP - CP)\mathrm{dc}\,\iota v,$$
or the equivalent
$$BP + AP = k(DP + CP)\mathrm{cd}\,\iota v.$$
The locus when u is a constant is given by
$$BP - AP = k(DP + CP)\mathrm{sn}\,u,$$
or $\qquad BP + AP = (DP - CP)\mathrm{ns}\,u.$

The curves in each case are bicircular quartics having A, B, C, D for foci. They are symmetrical about both axes.

The curves $v = \text{const.}$ are found to be a series of ovals enclosing the points $(\pm 1, 0)$ but not the points
$$\left(\pm\frac{1}{k}, 0\right).$$
The ends of the axes of these ovals are the points
$$(\pm\mathrm{cd}\,\iota v, 0) \quad \text{and} \quad (0, \pm\iota\,\mathrm{sn}\,\iota v).$$
When v is indefinitely diminished the oval shrinks up into the straight line between A and B. As v increases in magnitude irrespective of sign the oval swells out. The points on the axis of x are points of undulation when $2\,\mathrm{cd}^2\iota v = 1 + 1/k^2$, and for greater values the oval swells out above and below the axis of x, and is narrowest at the axis. In the limit when $v = \pm K'$, it becomes the part of the axis of x beyond $(\pm 1/k, 0)$, together with the line at infinity.

The curves $u = \text{const.}$ consist each of a pair of ovals, one enclosing the points $(1, 0)(1/k, 0)$ the other the points $(-1, 0)(-1/k, 0)$. Each of these cuts each of the curves $v = \text{const.}$ orthogonally.

Of the two ovals, the one on the positive side of the axis of y belongs to the values u and $2K - u$ (u being positive) and the other to the values $-u$ and $-2K + u$.

When $u = \pm K$ the corresponding oval shrinks into the straight line between $(\pm 1, 0)$ and $(\pm 1/k, 0)$, the upper or lower sign being taken throughout. When $u=0$ the oval swells out until it becomes the axis of y with the line at infinity.

The curve $v = \tfrac{1}{2}K'$ is the circle whose centre is the origin and radius $k^{-\frac{1}{2}}$.

§ 132. Since
$$\operatorname{dn}(u+\iota v, k) = k' \operatorname{sn}(v - \iota u + K' - \iota K, k'),$$
the figures for the function dn will be of the same general nature as those for sn. The foci
$$(\pm 1, 0)(\pm 1/k, 0)$$
are replaced by $\quad (\pm k', 0)(\pm 1, 0)$
respectively, and the single central ovals are now the curves $u = \text{const.}$, the pairs of ovals belonging to the system $v = \text{const.}$ The curve $u = \tfrac{1}{2}K$ is a circle of radius $k'^{\frac{1}{2}}$.

In the case of the function cn the figures are different.

Putting $x + \iota y = \operatorname{cn}(u + \iota v)$, we have
$$x = \operatorname{cn} u \operatorname{cn} \iota v / (1 - k^2 \operatorname{sn}^2 u \operatorname{sn}^2 \iota v),$$
$$y = \iota \operatorname{sn} u \operatorname{sn} \iota v \operatorname{dn} u \operatorname{dn} \iota v / (1 - k^2 \operatorname{sn}^2 u \operatorname{sn}^2 \iota v).$$

The curves $u = \text{const.}$, $v = \text{const.}$ are still bicircular quartics but the four real foci are not collinear. They are the points $(\pm 1, 0)(0, \pm k'/k)$, each of these pairs being collinear with the antipoints of the other.*

Each of the curves consists of a single oval. The curves $u = \text{const.}$ enclose the foci $(0, \pm k'/k)$ and not $(\pm 1, 0)$. The curve $u = 0$ consists of the parts of the axis of x beyond the points $(\pm 1, 0)$, the curve $u = \pm K$

---

* This may be compared with § 131 by means of the formula
$$\operatorname{cn}(u, k) = \operatorname{sn}(k'K - k'u, \iota k/k'),$$
which follows from equations (20) of § 26.

of the line between the points $(0, \pm k'/k)$. As u decreases numerically from $\pm K$ to 0, or increases from $\pm K$ to $\pm 2K$, the oval swells out. It has points of undulation on the axis of x when

$$2\,\mathrm{cn}^2 u = 1 - k'^2/k^2 \quad \text{if} \quad k^2 > k'^2.$$

When $\mathrm{cn}^2 u$ is greater than the value thus given the oval is shaped rather like a dumb-bell, and the two ends of it expand to infinity as u diminishes to 0 or increases numerically to $\pm 2K$.

Since

$$k\,\mathrm{cn}(u+\iota v, k) = -\iota k'\,\mathrm{cn}(v - \iota u + K' - \iota K, k'),$$

the general form of the curves $u = \text{const.}$, $v = \text{const.}$ is the same if one set is turned through a right angle. There will be points of undulation on one of the curves $v = \text{const.}$ if $k'^2 > k^2$, that is if there are not on any of the curves $u = \text{const.}$

§ 133. These bicircular quartics are shown in figures $4a, 5a, 6a$, for sn, cn, dn respectively. They have been drawn to scale with some care for the value $\sqrt{2}-1$ of k, and for values of u and v which are successive multiples of $\tfrac{1}{4}K$ and $\tfrac{1}{4}K'$ respectively.

In each case the curves $u = \text{const.}$ are drawn thick, and the curves $v = \text{const.}$ thin. The figures $4b, 5b, 6b$ show on the same scale the corresponding variations in the argument, corresponding lines in the two figures being numbered alike. Only one period-parallelogram has been drawn for each function. In each case the centre is at the origin.

The figures $4b, 5b, 6b$ are reproduced on a smaller scale as $4c, 5c, 6c$ the parallelograms being divided into the regions that correspond respectively to the four quadrants in $4a, 5a, 6a$.

In figure $6a$ the curves $v = 0$, $v = \pm \tfrac{1}{4}K'$, $v = \pm \tfrac{7}{4}K'$, $v = \pm 2K'$ are too small to be shown.

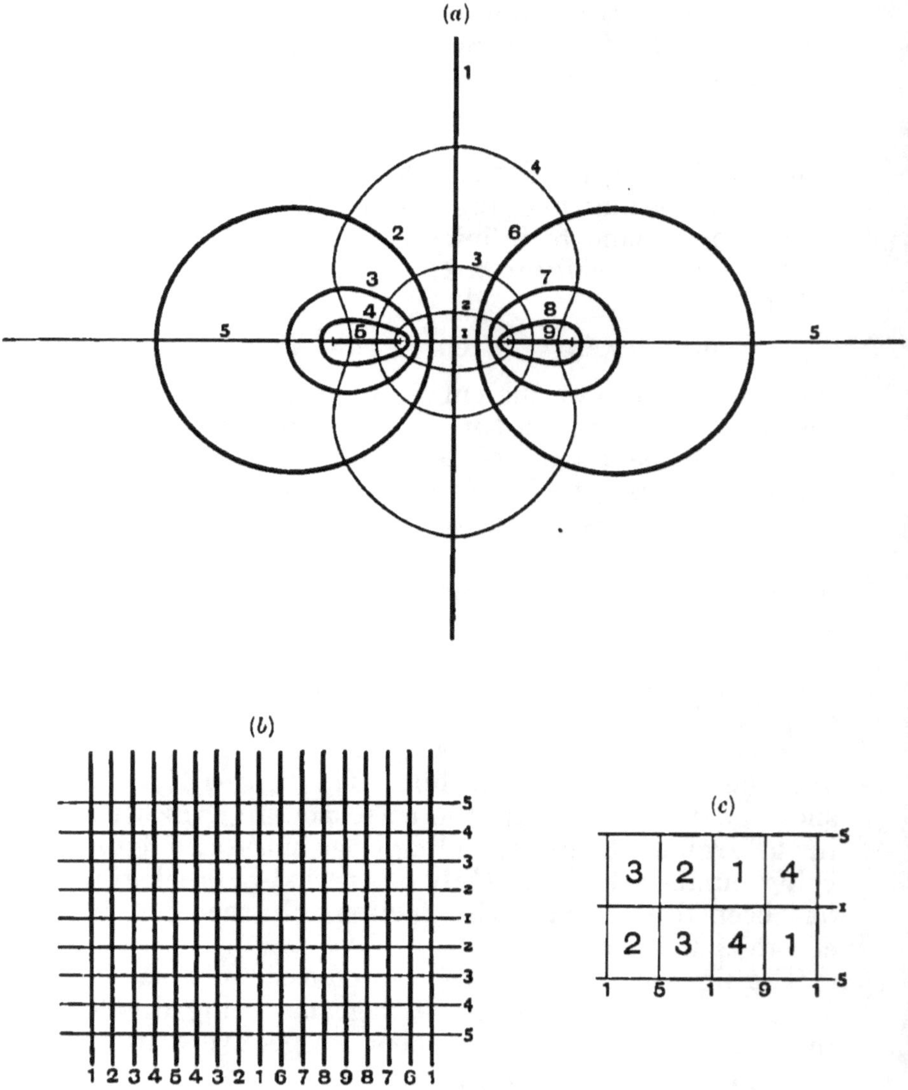

GRAPHICAL REPRESENTATION.

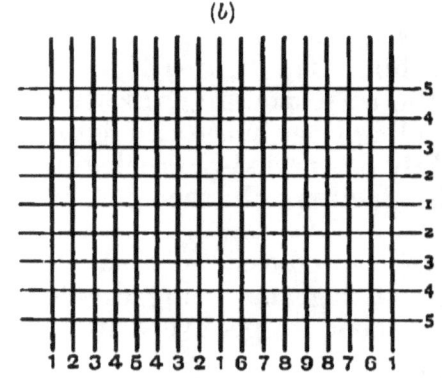

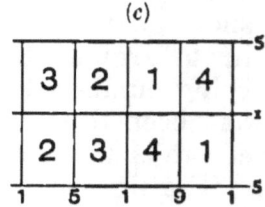

Fig. 4.

134   ELLIPTIC FUNCTIONS.

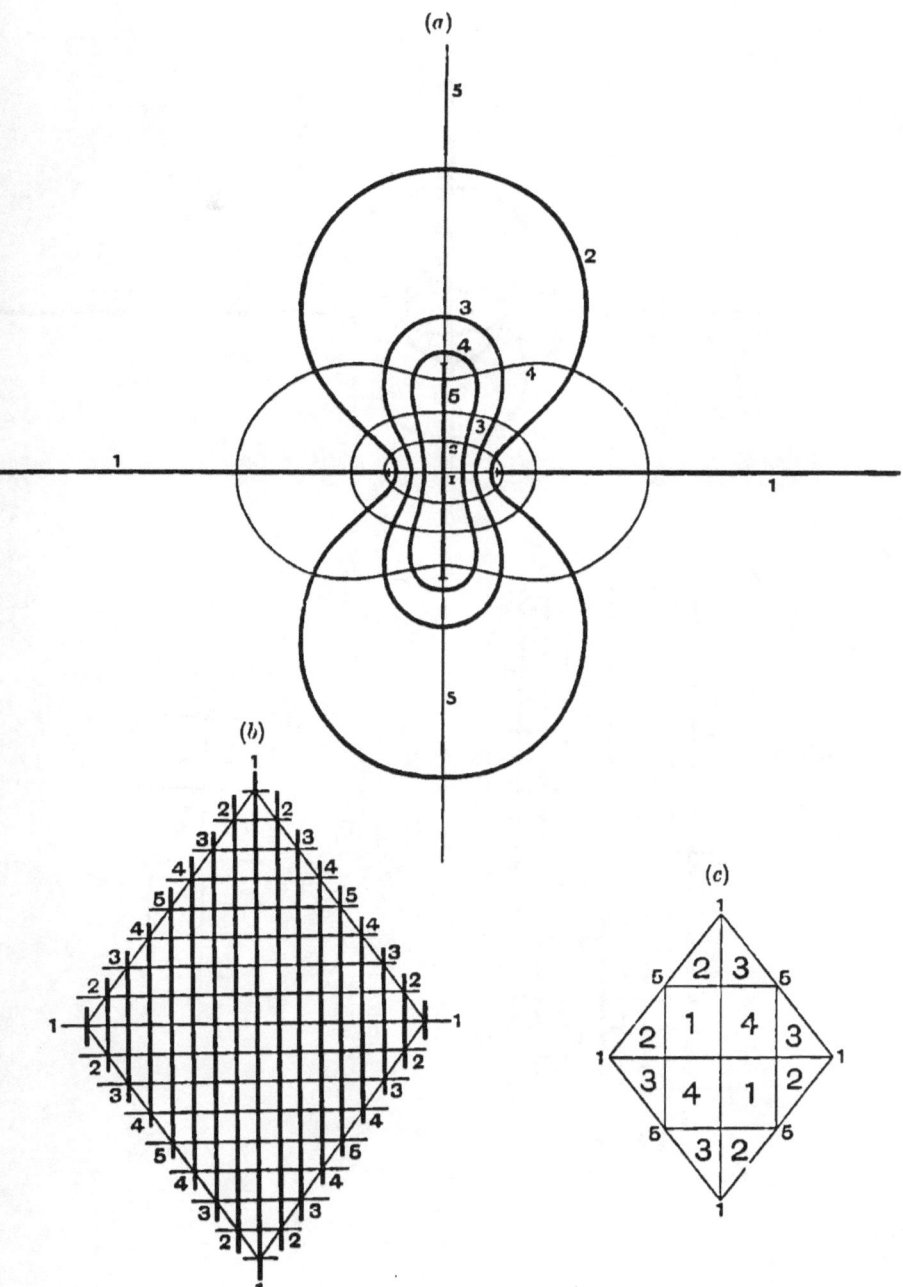

GRAPHICAL REPRESENTATION.

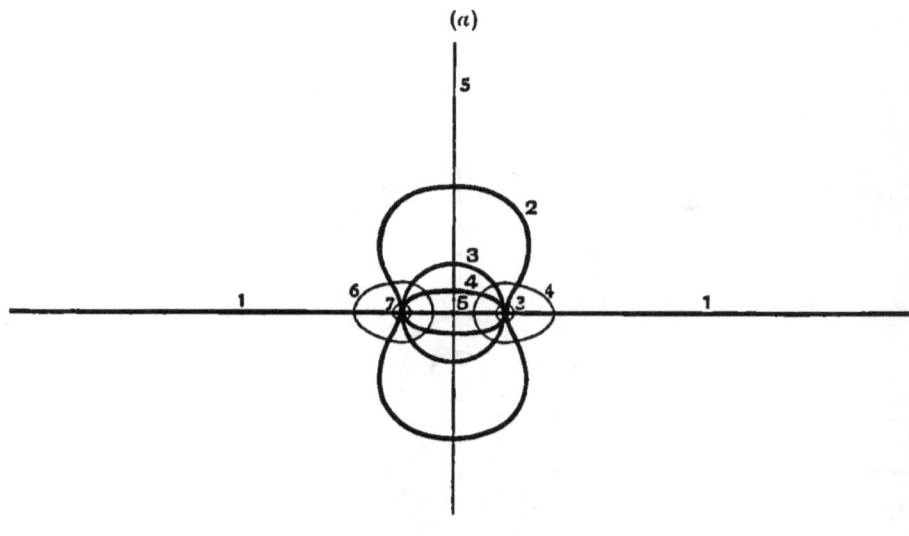

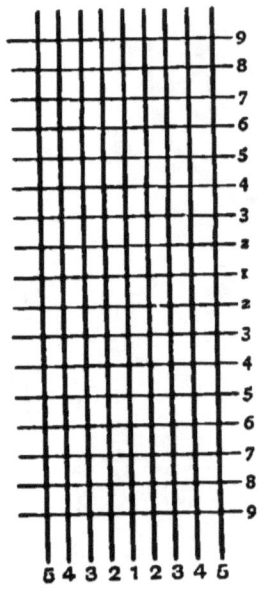
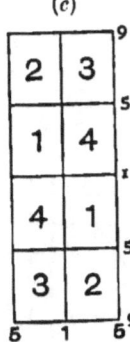

Fig. 6.

APPENDIX B.

HISTORY OF THE NOTATION OF THE SUBJECT.

§ 134. The notation used by Legendre was as follows:—*

$$F(k, \theta) = \int_0^\theta (1 - k^2 \sin^2\theta)^{-\frac{1}{2}} d\theta,$$

$$E(k, \theta) = \int_0^\theta (1 - k^2 \sin^2\theta)^{\frac{1}{2}} d\theta,$$

$$F(k, \tfrac{1}{2}\pi) = F_1(k), \ E(k, \tfrac{1}{2}\pi) = E_1(k),$$

$$\Pi(k, n, \theta) = \int_0^\theta (1 - k^2 \sin^2\theta)^{-\frac{1}{2}} d\theta / (1 + n \sin^2\theta),$$

$$\Delta\theta = (1 - k^2 \sin^2\theta)^{\frac{1}{2}}.$$

Jacobi and Abel proposed to take $F(k, \theta)$ as the independent variable. Putting u for this, Jacobi called θ the amplitude of u, or shortly am u. Then $\sin\theta$, $\cos\theta$, $\Delta\theta$ were the sine, cosine, and Δ of the amplitude of u, or as he wrote them,

$$\sin \operatorname{am} u, \quad \cos \operatorname{am} u, \quad \Delta \operatorname{am} u.$$

* The expressions $F(k, \theta)$, $E(k, \theta)$, $\Pi(k, n, \theta)$, were called the First, Second, and Third Elliptic Integrals respectively.

HISTORY. 137

He used the symbol coam u for am$(K-u)$, and also tan am u, sin coam u, etc.

He changed the meaning of the symbols E, Π to those we have given (Chap. V.), and also brought in the function Z.

It was proposed by Gudermann to write sn, cn, dn for sin am, cos am, Δ am, and the notation sc, cd, etc., was introduced by Dr. Glaisher. Sometimes tn is written for sc, ctn for cs. The function gd (see §75, *note*) is the amplitude, the modulus being unity. For the notation of Weierstrass see Chap. VII.

In the further development of the subject other symbols are wanted. Jacobi used the Greek capitals Θ and H; the functions Θu, Hu may be defined as follows:—

$$\Theta u = \exp\left(\int^u Zv\, dv\right),$$

$$\mathrm{H}u = \sqrt{k}\,.\,\Theta u\,.\,\operatorname{sn} u.$$

The arbitrary constant in the value of Θ is not determined until a later stage.

Some of the properties of the function Θu have been suggested in the examples to Chapter VI.

MISCELLANEOUS EXAMPLES

(FROM EXAMINATION PAPERS).

1. Prove that
$$\operatorname{sd}(x+y)\operatorname{sd}(x-y) = \frac{\operatorname{sd}^2 x - \operatorname{sd}^2 y}{1 + k^2 k'^2 \operatorname{sd}^2 x \operatorname{sd}^2 y},$$

$$\operatorname{cn}(x+y) = \frac{\operatorname{sd} x \operatorname{cn} x - \operatorname{sd} y \operatorname{cn} y}{\operatorname{sd} x \operatorname{cn} y - \operatorname{sd} y \operatorname{cn} x}.$$

2. Show that
$$\operatorname{sn} a \operatorname{sn} \beta = \frac{\operatorname{cn} a \operatorname{cn} \beta - \operatorname{cn}(a+\beta)}{\operatorname{dn}(a+\beta)} = \frac{\operatorname{dn} a \operatorname{dn} \beta - \operatorname{dn}(a+\beta)}{k^2 \operatorname{cn}(a+\beta)}.$$

3. Two sets of orthogonal curves (Cartesian ovals) being defined by the equation
$$x + \iota y = \operatorname{sn}^2\{\tfrac{1}{2}(u + \iota v), k\},$$
show that the polar coordinates of any point (u, v) are given by
$$\cos \theta = \frac{-\operatorname{cn}(u, k) + \operatorname{dn}(u, k)\operatorname{dn}(v, k')}{\operatorname{dn}(u, k) - \operatorname{cn}(u, k)\operatorname{dn}(v, k')},$$

$$\sin \theta = \frac{k'^2 \operatorname{sn}(u, k) \operatorname{sn}(v, k')}{\operatorname{dn}(u, k) - \operatorname{cn}(u, k)\operatorname{dn}(v, k')},$$

$$r = \frac{1 - \operatorname{cn}(u, k)\operatorname{cn}(v, k')}{\operatorname{dn}(v, k') + \operatorname{dn}(u, k)\operatorname{cn}(v, k')}.$$

MISCELLANEOUS EXAMPLES. 139

4. Prove that the functions
$$(\operatorname{cs} u \operatorname{cd} u \operatorname{cn} u - k'^2 \operatorname{sc} u \operatorname{sd} u \operatorname{sn} u)^2$$
and
$$(\operatorname{ds} u \operatorname{dc} u \operatorname{dn} u + k^2 k'^2 \operatorname{sc} u \operatorname{sd} u \operatorname{sn} u)^2$$
have periods K and $\iota K'$.

5. If
$$u = \frac{1}{2}\int_\epsilon^\infty \left\{\frac{a-c}{(a+\lambda)(b+\lambda)(c+\lambda)}\right\}^{\frac{1}{2}} d\lambda,$$
where a, b, c are positive quantities in descending order of magnitude, then
$$\epsilon \operatorname{sn}^2 u = a \operatorname{cn}^2 u - c,$$
the modulus being $\{(a-b)/(a-c)\}^{\frac{1}{2}}$.

6. Show that
$$k^2 \operatorname{sn} \tfrac{1}{2}(u_1+u_2+u_3+u_4) \operatorname{sn} \tfrac{1}{2}(u_1+u_2-u_3-u_4)$$
$$\times \operatorname{sn} \tfrac{1}{2}(u_1-u_2+u_3-u_4)\operatorname{sn} \tfrac{1}{2}(u_1-u_2-u_3+u_4)$$
$$= \frac{d_1 d_2 d_3 d_4 - k^2 c_1 c_2 c_3 c_4 + k^2 k'^2 s_1 s_2 s_3 s_4 - k'^2}{d_1 d_2 d_3 d_4 - k^2 c_1 c_2 c_3 c_4 - k^2 k'^2 s_1 s_2 s_3 s_4 + k'^2}.$$

7. Show that the form assumed by a uniform chain of given length whose ends are at two fixed points is represented by the equation
$$k'^2 y = 2kb \operatorname{sn} \frac{x+c}{b},$$
when its moment of inertia about the axis of x has a stationary value.

8. Prove that
$$\operatorname{cn}(B-C)\operatorname{sn}(C-A) + \operatorname{dn}(B-C)\operatorname{sn}(A-B)$$
$$+ \operatorname{sn}(B-C)\operatorname{cn}(C-A)\operatorname{dn}(A-B) = 0.$$

9. Verify that
$$\frac{\{1-k^2\operatorname{sn}^2(c+d)\operatorname{sn}^2(a-b)\}\{1-k^2\operatorname{sn}^2(a+b)\operatorname{sn}^2(c-d)\}}{\{1-k^2\operatorname{sn}^2(a+b)\operatorname{sn}^2(a-b)\}\{1-k^2\operatorname{sn}^2(c+d)\operatorname{sn}^2(c-d)\}}$$
is a symmetric function of a, b, c, d.

ELLIPTIC FUNCTIONS.

10. Prove that
$$-\iota k^{\frac{1}{2}}\operatorname{sn}(u+\tfrac{1}{2}\iota K') = \frac{cd - \iota(1+k)s}{1+ks^2} = \frac{1+ks^2}{cd+\iota(1+k)s}$$
$$= \frac{d - \iota ksc}{c + \iota sd} = \frac{c - \iota sd}{d + \iota ksc},$$

where s, c, d denote $\operatorname{sn} u$, $\operatorname{cn} u$, $\operatorname{dn} u$ respectively.

11. Prove that
$$-k\operatorname{sn}^2(u+\tfrac{1}{2}\iota K') = \frac{D - \iota kS}{C + \iota S} = \frac{C - \iota S}{D + \iota kS}$$
$$= \frac{C - kD - k'^2 S}{D - kC} = \frac{D - kC}{C - kD + k'^2 S},$$

where $\quad S = \operatorname{sn} 2u, \quad C = \operatorname{cn} 2u, \quad D = \operatorname{dn} 2u$.

12. If $x_{\lambda\mu}$ denote $\operatorname{sc}(u_\lambda - u_\mu)\operatorname{cs}(u_\lambda + u_\mu)$ then
$$x_{41}x_{42}x_{43}x_{12}x_{23}x_{31} + x_{41}x_{23} + x_{42}x_{31} + x_{43}x_{12} = 0.$$

13. If $k^2 = -\omega$ (where $\omega^2 + \omega + 1 = 0$) then
$$\frac{1 - \operatorname{sn}(\omega - \omega^2)u}{1 + \operatorname{sn}(\omega - \omega^2)u} = \frac{1 - \operatorname{sn} u}{1 + \operatorname{sn} u}\left(\frac{1 - \omega \operatorname{sn} u}{1 + \omega \operatorname{sn} u}\right)^2.$$

14. If $\quad Qu = \wp u + \wp(u+\omega) - \wp\omega,$
then $(Q'u)^2 = 4Q^3u + 4(g_2 - 15\wp^2\omega)Qu - 14g_2\wp\omega - 22g_3$.

15. Evaluate $\int(\wp u - \wp v)^2 du$, and express $\int(\wp u - \wp v)^{-2}du$ in terms of $\int(\wp u - \wp v)^{-1}du$.

16. Find $\int \operatorname{nd} u\, du$.

Prove that
$$3\int \operatorname{dn}^4 u\, du = 2(1+k'^2)Eu + k^2 \operatorname{sn} u \operatorname{cn} u \operatorname{dn} u - k'^2 u,$$
$$k'^2\int \frac{\operatorname{sn} u\, du}{1 + \operatorname{sn} u} = E(u + K + \iota K') + \operatorname{dc} u,$$
$$2k\int_0^K \operatorname{sn} u\, du = \log\frac{1+k}{1-k}.$$

MISCELLANEOUS EXAMPLES.

17. Show that

$$\int_0^K \log \operatorname{sn} u \, du = -\tfrac{1}{4}\pi K' - \tfrac{1}{2}K \log k,$$

$$\int_0^K \log \operatorname{cn} u \, du = -\tfrac{1}{4}\pi K' + \tfrac{1}{2}K \log \tfrac{k'}{k},$$

$$\int_0^K \log \operatorname{dn} u \, du = \tfrac{1}{2}K \log k'.$$

(In the first put am $u = \theta$ and expand in powers of k.)

18. Prove the formulae

$$\int_0^K \log(1 - k^2 \operatorname{sn}^4 u) du = -\tfrac{1}{4}\pi K' + K \log \tfrac{2k'}{k^{\frac{1}{2}}},$$

$$\int_0^K \log(1 + \operatorname{dn} u) du = \tfrac{1}{4}\pi K' + \tfrac{1}{2}K \log k.$$

19. Prove that

$$\Pi(u, a) + \Pi(v, a) - \Pi(u+v, a)$$
$$= \tfrac{1}{4} \log \frac{\{1 - k^2 \operatorname{sn}^2(u+a)\operatorname{sn}^2(v+a)\}\{1 - k^2 \operatorname{sn}^2 a \operatorname{sn}^2(u+v-a)\}}{\{1 - k^2 \operatorname{sn}^2(u-a)\operatorname{sn}^2(v-a)\}\{1 - k^2 \operatorname{sn}^2 a \operatorname{sn}^2(u+v+a)\}}.$$

20. Expand $\int_0^K \operatorname{sn}^n u \, du$ in ascending powers of k^2; and thence, or otherwise, prove that

$$\frac{d}{dk}\int_0^K \operatorname{sn}^n u \, du = k \int_0^K \frac{\operatorname{sn}^{n+2} u}{\operatorname{dn}^2 u} du.$$

(Compare Ex. 12, Chap. IX.)

21. In Weierstrass' notation, if J is the absolute invariant as given by the equations

$$\frac{J-1}{27 g_3^2} = \frac{J}{g_2^3} = \frac{1}{\Delta},$$

142 ELLIPTIC FUNCTIONS.

then the periods satisfy the differential equation

$$J(1-J)\frac{d^2}{dJ^2}(y\Delta^{\frac{1}{12}}) + \tfrac{1}{6}(4-7J)\frac{d}{dJ}(y\Delta^{\frac{1}{12}}) = \tfrac{1}{144}y\Delta^{\frac{1}{12}}.$$

22. Verify that the expression of Ex. 19 agrees with that of Ex. 15, Chap. VI., and with that of § 67.

23. Find expressions for the arcs of the curves,

$$k'^2 y = 2kb \operatorname{sn} \frac{x+c}{b},$$

$$y = bk' \operatorname{nc} \frac{2x+c}{b}.$$

www.ingramcontent.com/pod-product-compliance
Lightning Source LLC
Chambersburg PA
CBHW030358170426
43202CB00010B/1420